Lecture Notes in Mechanical Engineering

Lecture Notes in Mechanical Engineering (LNME) publishes the latest developments in Mechanical Engineering—quickly, informally and with high quality. Original research or contributions reported in proceedings and post-proceedings represents the core of LNME. Volumes published in LNME embrace all aspects, subfields and new challenges of mechanical engineering.

To submit a proposal or request further information, please contact the Springer Editor of your location:

Europe, USA, Africa: Leontina Di Cecco at Leontina.dicecco@springer.com
China: Ella Zhang at ella.zhang@cn.springernature.com
India, Rest of Asia, Australia, New Zealand: Swati Meherishi at swati.meherishi@springer.com

Topics in the series include:

- Engineering Design
- Machinery and Machine Elements
- Mechanical Structures and Stress Analysis
- Automotive Engineering
- Engine Technology
- Aerospace Technology and Astronautics
- Nanotechnology and Microengineering
- Control, Robotics, Mechatronics
- MEMS
- Theoretical and Applied Mechanics
- Dynamical Systems, Control
- Fluid Mechanics
- Engineering Thermodynamics, Heat and Mass Transfer
- Manufacturing Engineering and Smart Manufacturing
- Precision Engineering, Instrumentation, Measurement
- Materials Engineering
- Tribology and Surface Technology

Indexed by SCOPUS, EI Compendex, and INSPEC.

All books published in the series are evaluated by Web of Science for the Conference Proceedings Citation Index (CPCI).

To submit a proposal for a monograph, please check our Springer Tracts in Mechanical Engineering at https://link.springer.com/bookseries/11693.

Giuseppe L. Conte · Olivier Sename
Editors

Proceedings of 12th International Conference on Mechatronics and Control Engineering

ICMCE 2024

Springer

Editors
Giuseppe L. Conte
Dipartimento di Ingegneria
dell'Informazione
Università Politecnica delle Marche
Ancona, Italy

Olivier Sename
CNRS, Grenoble INP
Université Grenoble Alpes
Grenoble, France

ISSN 2195-4356 ISSN 2195-4364 (electronic)
Lecture Notes in Mechanical Engineering
ISBN 978-981-96-6451-1 ISBN 978-981-96-6452-8 (eBook)
https://doi.org/10.1007/978-981-96-6452-8

This Springer imprint is published by the registered company Springer Nature Singapore Pte Ltd.
The registered company address is: 152 Beach Road, #21-01/04 Gateway East, Singapore 189721, Singapore

Committees

International Advisory Committee

Graziano Chesi, The University of Hong Kong, Hong Kong
József Bokor, Institute for Computer Science and Control (SZTAKI), Hungary
Pierre Borne, IEEE Life Fellow, Ecole Centrale de Lille, France

Conference General Chair

Giuseppe Conte, Università Politecnica delle Marche, Italy

Conference Co-chair

Olivier Sename, Grenoble Institute of Technology, France

Program Chairs

Péter Gáspár, Budapest University of Technology and Economics, Hungary
Ricardo Ambrocio Ramirez-Mendoza, Tecnológico de Monterrey (ITESM), Mexico

Program Co-chairs

Elena Zattoni, Università di Bologna, Italy
Sallehuddin Bin Mohamed Haris, Universiti Kebangsaan Malaysia, Malaysia

Publicity Chairs

Ke-Lin Du, Concordia University, Canada
Sergei Alexandrov, RUDN University, Moscow, Russian Federation

Technical Committees

Dario Mendoza Chipantasi, Universidad de las Fuerzas Armadas ESPE, Ecuador
David Scaradozzi, Università Politecnica delle Marche, Italy
Emin Taner Elmas, Igdir University, Türkiye
Filippo d'Ippolito, Università di Palermo, Italy
Garje Channabasappa Mohan Kumar, National Institute of Technology Karnataka, India
Gokhan Erdemir, The University of Tennessee at Chattanooga, USA
Jaharah Binti A. Ghani, Universiti Kebangsaan Malaysia, Malaysia
K. Kadirgama, Universiti Malaysia Pahang, Malaysia
Karina R. G. Silva, Federal University of Goias, Brazil
Mehmet Emir Koksal, Ondokuz Mayis University, Türkiye
Mingbiao Yu, Southeast University, China
Omar Y. Ismael, Ninevah University, Iraq
Recai Kuş, Selcuk University, Türkiye
Sarunya Promkotra, Khon Kaen University, Thailand
Silvia Maria Zanoli, Universitrà Politecnica delle Marche, Italy
Viswanathan Kodakkal Kannan, University Teknologi Malaysia, Malaysia
Zareena Kausar, Air University, Pakistan
Zeki Yagiz Bayraktaroglu, Istanbul Technical University, Türkiye

Preface

ICMCE is an annual research conference aiming to provide a platform for engineers and scientists to present their ideas and experiences in the fields of Mechatronics and Control Engineering. The 12th International Conference on Mechatronics and Control Engineering (ICMCE 2024) was held in Budapest, Hungary during January 25–27, 2024.

The accepted papers were peer-reviewed by reviewers drawn from the committees, external reviewers, and editorial board depending on the subject of the papers. Efforts taken by peer reviewers provided constructive critical comments; improvements and corrections to the authors are gratefully appreciated. The appreciation also goes to other conference committee members, keynote/invited speakers, session chairs and all the authors, thanks for their understanding, support and efforts. Without their help, ICMCE couldn't be held successfully.

Presentations of the selected papers, grouped into three sessions. This year, the program committee organized a technical tour to the Innovation and Demonstration Space at SZTAKI the Institute for Computer Science and Control (Budapest). We are very thankful to the Institute staff for this welcome. We are really pleased to share the program with around 30 attendees as well as with the entire community through the proceedings. We are glad that experts in the various fields of Mechatronics and Control Engineering are presented and discussed their ideas and latest development in the field.

Hope all the participants had wonderful time during conference and got fruitful inspirations from the presentations delivered by speakers and authors. Hope to see you next year again!

Ancona, Italy
Grenoble, France

Giuseppe L. Conte
Olivier Sename
Conference Chairs

Contents

About the Editors

Giuseppe L. Conte received the Laurea (Hons.) degree in Mathematics from the University of Genoa, Genoa, Italy, in 1974. He has been Professor of Automation with the Polytechnic University of Marche, Ancona, Italy. He authored more than 300 papers in journals, conference proceedings, and books; he coedited four books and coauthored the research monographs *Algebraic Methods for Nonlinear Control Systems Theory and Applications* (Springer Verlag, 2007). His research interests are in algebraic and geometric system and control theory and in robotics. Professor Conte served in the editorial boards of several journals and as Program Chair and General Chair of systems and control conferences. He was Chairman of the IFAC TC on Linear Control Systems and Chairman of the Italian Chapter of the IEEE CSS.

Olivier Sename received the Ph.D. degree in Automatic Control from École Centrale de Nantes, Nantes France, in 1994. He is currently Professor with the Institut Polytechnique de Grenoble, GIPSA-lab, Grenoble, France. His main research interests include linear parameter varying systems and automotive applications. He is currently Co-author of two books, 60 international journal papers, and more than 200 international conference papers. Dr. Sename was the General Chair of the IFAC Joint Conference SSSC-TDS-FDA 2013, of the 1st IFAC Workshop on Linear Parameter Varying Systems 2015, and he was IPC Chair of the 2nd IFAC Workshop LPVS 2018. He has led several industrial (Renault, Volvo Trucks, JTEKT, Delphi) and international (Mexico, Italy, Hungary) collaboration projects.

Performance Analysis and Processing of Functional Materials

Influence of Graphene Coating on Electrical Discharge Machining (EDM) Utilizing a Graphite Electrode

Israa Dheyaa Khalaf Alrubaye, Gualtiero Fantoni, Davide Melis, Massimo Arcioni, and Dario Clori

1 Introduction

In recent years, there has been a growing demand for diverse tool electrodes in the electrical discharge machining field. These requirements include higher speed, improved efficiency, enhanced performance stability, extended tool lifespan, and cost reduction, even when operating under more challenging conditions within a typical environment. To address these demands, the development of coated tools has emerged as a solution. Coated tools involve the application of protective layers onto the tool's surface to enhance its performance and durability. These coatings provide some advantages such as a higher material removal rate, resistance to wear, and thermal stability, enabling the tools to operate more effectively in demanding machining applications. The invention of coated tools has significantly contributed to meeting the evolving needs of the machining industry [1, 2].

Many materials could be used for coating electrodes of EDM, such as silicon (Si), aluminum (Al), chromium (Cr), copper (Cu), graphene (Gr), and carbon nanotube (CNT). Carbon forms (graphene and carbon nanotubes) are the most suitable materials due to their superior properties for improving the machining process.

Carbon, with its versatile forms and structures, has been extensively utilized in various technological applications and has had a significant impact on human life for centuries. The diverse arrangements of carbon atoms give rise to distinct carbon

I. D. K. Alrubaye (✉)
Department of Smart Industry, University of Pisa, Pisa, Largo Lucio Lazzarino, Italy
e-mail: israa.alrubaye@phd.unipi.it

G. Fantoni
Department of Civil and Industrial Engineering, University of Pisa, Pisa, Largo Lucio Lazzarino, Italy

D. Melis · M. Arcioni · D. Clori
Baker Hughes Nuovo Pignone Tecnologie S.r.l., Firenze, Italy

G. L. Conte and O. Sename (eds.), *Proceedings of 12th International Conference on Mechatronics and Control Engineering*, Lecture Notes in Mechanical Engineering,
https://doi.org/10.1007/978-981-96-6452-8_1

shapes, including graphite, graphene, and carbon nanotubes (CNTs). Each of these carbon structures possesses unique properties that set them apart from one another [3].

Graphene, an allotrope of carbon, represents a distinctive form of carbon material with remarkable properties. Extensive research has revealed that graphene possesses exceptional strength, estimated to be approximately 300 times greater than that of steel, making it one of the strongest known materials. Moreover, its intrinsic hardness surpasses that of diamonds, which are renowned for their exceptional hardness. Stacked layers of graphene constitute graphite, a common material known for its lubricating properties and use in pencils. When graphene sheets are rolled up, they form CNTs. CNTs are tubular, comprising carbon atoms arranged in a cylindrical structure. They can be classified into two main categories: single-walled and multi-walled carbon nanotubes [4–6].

Graphene also has garnered significant attention in the scientific community due to its potential to provide corrosion resistance when applied as coatings. Therefore, this work will employ graphene-coated electrodes for the finishing and roughing phases of die-sinking EDM technology. The coated electrodes will be compared with non-coated electrodes to investigate differences in terms of efficiency, if any, in the EDM process.

2 State of the Art

Coating technology has been used widely in the last years by many researchers to enhance the EDM process and increase tool life. Different materials have been used for coating as follows:

In [7], the author investigated the utilization of nickel-coated composite electrodes and diamond-nickel-coated composite electrodes for the machining of die steel.

In [8], the author examined the impact of aluminum–chromium–nickel-coated aluminum electrodes.

In [9], the author used an electrode consisting of a copper substrate with a coating of nichroloy. In [10], the author used an aluminum–chromium–nitride-coated aluminum electrode. In [11], the author used a nickel-coated aluminum electrode.

In [12], the author employed a normal brass tool, a tin-coated brass tool, and a cryogenically treated brass tool.

In [13], the author used a titanium nitride-coated tungsten carbide electrode.

Most studies used different types of materials, but very little research has been done for using carbon forms such as graphene or CNT for coating electrodes, some of these studies are listed below:

In [14], the author used copper single-wall carbon nanotube (Cu-SWCNT) composite coatings applied on 6061 aluminum to machine AISI 304 stainless steel using EDM oil as the dielectric fluid. The results demonstrated an improvement in machining parameters with composite-coated 6061 aluminum electrodes compared to uncoated or base 6061 aluminum electrodes. Specifically, the Material Removal

Rate (MRR) increased by 14.5%, and the electrode wear rate (EWR) decreased by 29.6% when Cu-SWCNT composite coatings were applied.

In [15], the author used graphene and carbon nanotubes as coating materials for brass electrodes during machining of Inconel 825. The results showed improvements in machining performance with coated brass electrodes compared to uncoated electrodes. Specifically, the graphene-coated electrodes showed an improvement of 16.9% in MRR, 25.6% in EWR, and 14.3% in terms of better surface roughness (smoother).

In [16], the author used a carbon-coated tungsten carbide electrode for the machining of titanium alloy. The carbon coating played a crucial role in increasing productivity and enhancing precision in micro-EDM processes. In [17], the author used graphene as a coating material for aluminum electrodes to machine titanium alloy. The results showed the graphene-coated increased MRR and reduced EWR. The coating reduced the formation of microcracks and pores, which were uniformly distributed across the surface.

In this work, graphite electrodes were coated with two types of graphene (pure graphene and graphene/zirconium) and used in both: roughing and finishing (EDM) operations. This investigation aims to evaluate the impact of graphene coatings on the EDM process and assess their potential for enhancing the electrode's performance.

3 Experimental Setup

To optimize the adhesion of the coating on the graphite electrode, a cleaning procedure was implemented to eliminate graphite dust. The coating process involved the utilization of both PVD (physical vapor deposition) and PECVD (plasma-enhanced chemical vapor deposition) combining techniques, which were performed by NANESA company. Two types of coatings, graphene/zirconium and graphene, were applied with 8–10 μm and 6–8 μm thicknesses, respectively. This experimental setup aimed to investigate the impact of the coating on both the finishing and roughing EDM phases. These trials have been carried out using an advanced die-sinking EDM machine (Agie Charmilles model). The experimental design consisted of several experiments with different cutting depths. Each experiment involved the use of three graphene electrode types: pure graphene coated, graphene/zirconium coated, and non-coated. Each electrode was used for machining a stainless-steel workpiece. The graphene/zirconium coating offers a combination of high electro-conductivity (provided by graphene), ionic conductivity, and high thermal stability (given by cubic zirconia). All electrodes have a usable length for the EDM process of 26.5 cm. The finishing process was performed with one side of the electrode, while the roughing process was done using the opposite side. Figure 1 shows the coated electrodes and the machine setup, while the experiment parameters are shown in Table 1.

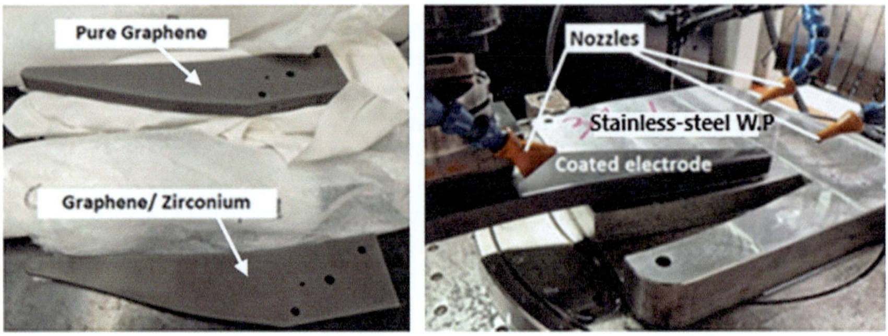

Fig. 1 The experimental setup. **a** Types of coated graphite electrodes. **b** Machine setup for roughing EDM process

Table 1 Variable and fixed parameters for the EDM process

Parameters	Value	
	Roughing	Finishing
Current (A)	140	6.3
Voltage (V)	100	100
Pulse on time (μs)	205	37
Pulse off time (μs)	24	21
Cutting depth (mm)	2, 4, 8, 14	0.3, 0.6, 0.9, 1.2, 1.5, 1.8, 2.1
Cutting length (cm)	13.5	
Dielectric fluid	EDM oil	
Electrode material type	Graphite	
Electrode coating types	Pure graphene and graphene zirconium	
Electrode shape	Vane shape, L (26.5 cm)	
Workpiece material	Stainless steel	

4 Results and Discussions

This study aimed to assess the effects of two different types of coatings on EDM performances. An optical microscope was used to investigate the difference in surface texture on the coated electrodes for roughing and finishing phases. Figure 2 shows the observed distinctions in the coated graphite electrode's sides, providing valuable insights into the impact of the coatings on the machining process using an optical microscope (Nikon) with micrometer scale (1e + 006), pixel (1638, 400), and 50% zoom resolution. In the context of the electrode surface quality, it is evident that the quality of the utilized side for the finishing process is better than that of the non-utilized part on the same side for both graphene and graphene/zirconium coating electrodes. Additionally, the non-utilized section of the graphene/zirconium-coated

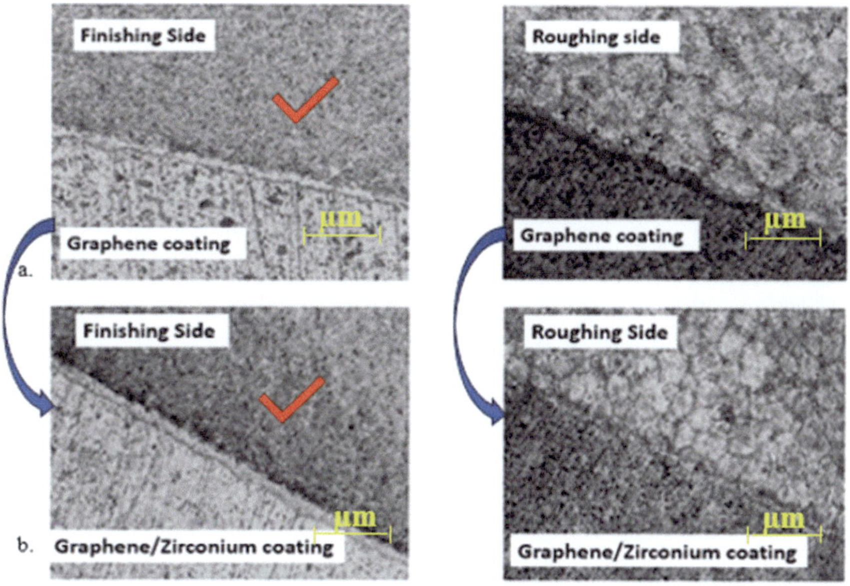

Fig. 2 Visual differences between electrode sides for finishing and roughing processes. **a** Graphene-coated electrode. **b** Graphene/zirconium-coated electrode

electrode exhibits better surface quality than the non-utilized section of the graphene-coated electrode for both finishing and roughing processes.

These experiments were supplemented by extensive measurements of EDM performance parameters, including machining time, electrode wear, and roughness value on machined surfaces, as covered in more detail in the following paragraphs.

4.1 Total Machining Time

One of the most important parameters to assess EDM process efficiency is the machining time. Its reduction enhances productivity, reduces costs, and optimizes machine utilization and workflow maintaining in this way a better control over the production quality of components. Table 2 shows the effect of the coated graphite electrode on the EDM process time (the numbers have been normalized).

In general, it is observed that the machining time associated with pure graphene and graphene/zirconium-coated graphite electrodes is lower compared to non-coated graphite electrodes in both roughing and finishing EDM phases as shown in Fig. 3a and 4a.

Table 2 Effect of graphene and graphene/zirconium-coated graphite electrode on time

Process type	Cutting depth (mm)	Total Machining time (min)		
		Non-coated electrode	Graphene-coated electrode	Graphene-zirconium-coated electrode
Roughing	2	1.00	**0.70**	**0.70**
	4	1.00	**0.70**	**0.70**
	8	2.60	**2.10**	**2.10**
	14	2.60	**2.80**	**2.80**
Finishing	0.3	1.00	**0.62**	0.65
	0.6	0.65	0.68	0.68
	0.9	0.71	**0.71**	**0.71**
	1.2	0.77	**0.74**	0.77
	1.5	0.78	**0.75**	**0.75**
	1.8	0.83	0.77	**0.75**
	2.1	0.80	0.75	**0.75**

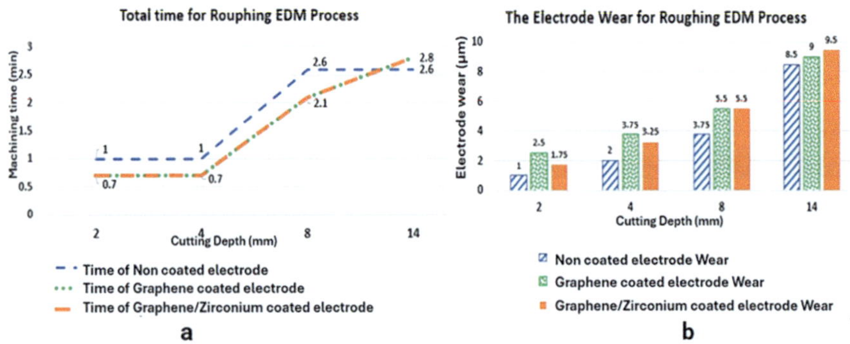

Fig. 3 Result analysis for roughing process. **a** Total machining time. **b** The electrode wears

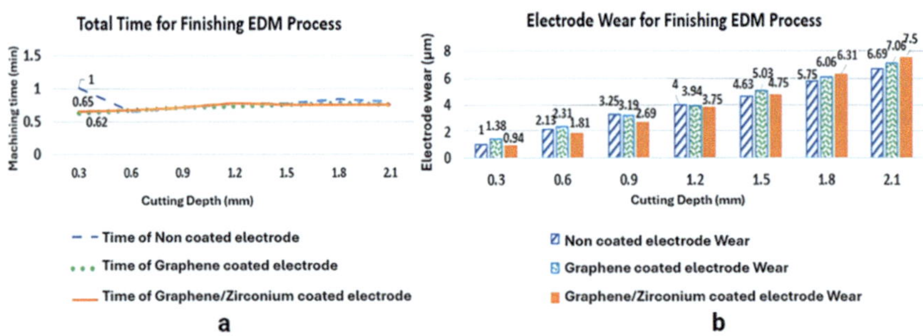

Fig. 4 Result analysis for finishing EDM process. **a** Total time. **b** Electrode wear

Table 3 Effect of graphene and graphene/zirconium coating on graphite electrode wear

Process type	Cutting depth (mm)	Electrode wear (μm)		
		Non-coated electrode	Graphene-coated electrode	Graphene-zirconium-coated electrode
Roughing	2	1.00	2.50	1.75
	4	2.00	3.75	3.25
	8	3.75	5.50	5.50
	14	8.50	9.00	9.50
Finishing	0.3	1.00	1.38	**0.94**
	0.6	2.13	2.31	**1.81**
	0.9	3.25	**3.19**	**2.69**
	1.2	4.00	**3.94**	**3.75**
	1.5	4.63	5.03	4.75
	1.8	5.75	6.06	6.31
	2.1	6.69	7.06	7.50

4.2 *Electrode Wear Measurement*

Electrode wear in the EDM process has a detrimental effect on process efficiency in multiple aspects, including dimensional accuracy, surface quality, electrode wear rate, and overall cost. Therefore, electrode wear reduction is certainly one of the fundamental requirements to keep EDM technology competitive with other material removal technologies. Table 3 shows the effect of coating on electrode wear (the numbers have been normalized).

A significant disparity in electrode wear is observed between the roughing and finishing EDM phase. During the roughing phase, it is evident that the wear is greater for both types of coated graphite electrodes compared to non-coated graphite electrodes. This disparity can be attributed to the use of more aggressive electrical parameters, typically employed in roughing operations. In the finishing process, it is observed that the use of graphene/zirconium-coated graphite electrodes showed the lowest wear for depth up to 0.9 mm when compared to all other electrodes, as shown in Fig. 3b and 4b. This finding suggests that the graphene/zirconium coating exhibits superior resistance to wear during the finishing process. In general, these observations highlight the impact of coating type and the variation in electrical parameters between roughing and finishing processes on electrode wear in EDM.

4.3 *Machined Surface Roughness*

In the EDM process, the component's surface quality plays a critical role, as it is an important requirement for the component specification in terms of performance and reliability. Several parameters, including current, pulse on time, pulse off time, voltage, dielectric fluid, and flushing type, can influence surface roughness. In this study, vane-shaped graphite electrodes were used for both graphene- and graphene/zirconium-coated and non-coated electrodes, for the roughing and finishing phases. The investigation focused on evaluating the impact of coated and non-coated electrodes concerning the roughness value obtained for both roughing and finishing phases. The findings related to the Ra values obtained are presented in Table 4 and Fig. 5.

Experimental results indicate that the utilization of graphene/zirconium-coated graphite electrode leads to the attainment of a smoother surface, with a lower Ra roughness value, for the finishing phase compared to all other electrodes. Conversely, for roughing operation, the graphene/zirconium coating seems to show the worst surface roughness value due to the high and aggressive electrical parameters (mainly for current) used in this experiment, which led to more aggressive electrode wear. This could be because the graphene/zirconium coating is quickly lost unevenly producing an electrode surface where the coating effect will be almost negligible if not worse than a non-coated electrode. Therefore, the process could be affected by the coating when using a high current that could lead to a higher surface roughness due to an uneven coating peel-off.

Table 4 Effect of graphene- and graphene/zirconium-coated graphite electrode on the machined surface roughness

Process type	Machined surface roughness (μm)		
	Non-coated electrode	Graphene-coated electrode	Graphene-zirconium-coated electrode
Roughing	(11.15–12.1) μm	(10.46–13.38) μm	(12.05–14.25) μm
Finishing	(3–3.2) μm	(3–3.2) μm	(**2.9–3**)) μm

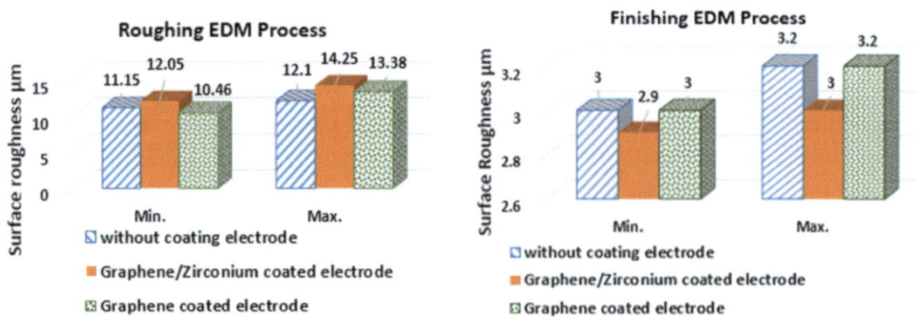

Fig. 5 Machined surface quality comparison among different electrodes for roughing and finishing EDM processes

5 Conclusion

In this study, the obtained conclusions shed light on the relative performance of graphene coatings, and their ramifications on distinct parameters of the EDM process, Figs. 3, 4 and 5. The parameters investigated include machining time, electrode wear, and workpiece surface roughness. The following conclusions are drawn:

- Currently, the utilization of graphene concerning coating applications is significantly less widespread compared with other materials for coating purposes. Besides the fact that no previous studies have specifically investigated the impact of graphene or graphene/zirconium-coated graphite electrodes to evaluate the performances of EDM technology.
- When considering the total machining time, which is also indicative of MRR capability, it is observed that both the graphene/zirconium-coated electrode and the graphene-coated electrode have similar performance in roughing processes. Coated electrodes achieved shorter erosion time than non-coated electrodes. The highest improvement was observed in the roughing process with a cutting depth of 8 mm producing approximately 19% time reduction for both coated electrodes compared to non-coated electrode. For the finishing operation, the highest improvement was obtained at 0.3 mm cutting depth with a graphene-coated electrode with a time reduction of just below 38% if compared with a non-coated electrode.
- Regarding electrode wear, it is found that the non-coated graphite electrode wears the least during the roughing process. This could be attributed to the interplay of thermal conductivity, flushing efficiency, and coating adhesion characteristics. These findings highlight the importance of electrode material selection and design considerations for optimizing roughing operations in EDM.
- For finishing processes, the graphene/zirconium-coated graphite electrode showed less wear than other types of electrodes in the initial erosion phase. The highest reduction in electrode wear was observed up to a cutting depth of 0.9 mm where the graphene/zirconium-coated graphite electrode had been aware of about 17% less than a non-coated electrode. However, as the erosion depth increased, this advantage was soon lost leading to rapid electrode wear with values for the graphene/zirconium-coated electrode higher than the non-coated as depicted in Fig. 4b.
- In roughing phase, looking at the minimum roughness values obtained across all electrodes, it was observed that the graphene-coated electrode achieved the lowest surface roughness with a value just above 10 microns. Conversely, when looking at the maximum roughness captured on the workpiece surface, with the same technological parameters used, the electrode without coating produced a surface roughness of about 12 microns (the lowest value among all electrodes). In the finishing phase, the graphene/zirconium-coated electrode produced the best surface with roughness values always lower than all other electrodes. Interestingly, the application of graphene coating on the graphite electrode does not significantly alter the surface roughness value when compared to a non-coated graphite electrode (3 μm and 3.2 μm seen in both), as shown in Fig. 5.

The current scientific literature reveals a scarcity of extensive investigations centered on graphene-coated electrodes, as evidenced by a limited number of scholarly publications devoted to exploring economically viable, practical, and commercially feasible methodologies for the large-scale synthesis of graphene. This lack of research findings can be attributed to the relatively recent integration of graphene into industrial contexts, which necessitates continued exploration and experimentation within the burgeoning field of EDM.

Based on these experiments, the graphene coating, as applied under these conditions, does not meet the demands of industry standards, especially when dealing with long machining times or harsh machining conditions. Additionally, there have been occasional concerns raised regarding potential health impacts associated with graphene coatings. Nevertheless, given the promising results obtained thus far, further studies are encouraged on graphene coatings in the EDM process. Experiments exploring different process conditions, coating composition, electrode interface adhesion improvements between electrode substrate and coating as well as increased coating thicknesses are therefore recommended.

Acknowledgements The authors acknowledge the facilities and the technical assistance of Dr. Emanuele Piciollo and Dr. Lorenzo Fabbri at (Nanesa Company, in Bucine, Arezzo, Italy), and Baker Hughes (Nuovo Pignone Tecnologie S.r.l) Company in Florence, Italy.

Funding This research is supported by the Ministry of University and Research (MUR) as part of the PON 2014–2020 "Research and Innovation" resources—Green/Innovation Action—DM MUR 1061/2022 with grant number: (I59J21017770008), and Baker Hughes (Nuovo Pignone Tecnologie S.r.l) Company.

References

1. Shirguppikar SS, Patil MS, Ganachari VS, Kolekar TV, Jadhav PS, Chougule AB (2018) Experimental investigation of CNT coated tools for EDM processes. Mater Today: Proc 5(2):7131–7140. https://doi.org/10.1016/j.matpr.2017.11.378
2. Alrubaye IDK, Fantoni G (2023) Toward green electrical discharge machining (EDM): state of art and outlook. Mach Sci Technol 27(1):63–105. https://doi.org/10.1080/10910344.2023.2194961
3. Salisu N, Mohd ZH, Zulkarnain Z, Nor AY (2018) Carbon-based nanomaterials/allotropes: a glimpse of their synthesis, properties and some applications. Materials 11(2):295. https://doi.org/10.3390/ma11020295
4. Urade AR, Lahiri I, Suresh KS (2023) Graphene properties, synthesis and applications: a review. Miner Metals Mater Soc 75(3):614–630. https://doi.org/10.1007/s11837-022-05505-8
5. Mai C, Hong H, Huang S (2012) Advantages of carbon nanotubes in electrical discharge machining. Int J Adv Manuf Technol 59(1–4):111–117. https://doi.org/10.1007/s00170-011-3476-2
6. Moshim GM, Anshuman K (2019) Effect of carbon Nano tube addition with dielectric fluid in die sink EDM. Int J Innov Technol Explor Eng 9(2):2278–3075. https://doi.org/10.35940/ijitee.B6333.129219

7. Liu Y, Wang W, Zhang W, Ma F, Yang D, Sha Z, Zhang S (2019) Experimental study on electrode wear of diamond-nickel coated electrode in EDM small hole machining. Adv Mater Sci Eng 2019(7):1–10. https://doi.org/10.1155/2019/7181237
8. Shirguppikar S, Patil MS, Phan NH, Muthuramalingam T, Dong PV, Tam NC, Tai BT, Minh ND, Duc NV (2021) Assessing the effects of uncoated and coated electrode on response variables in electrical discharge machining for Ti-6Al-4V titanium alloy. Tribol Ind 43(4):524–534. https://doi.org/10.24874/ti.1020.12.20.03
9. Baral SK, Pallavi P, Utkarshaa A (2021) Effect of nichroloy coated electrode on machining performance of electrical discharge machining using Inconel 625. Mater Today: Proc 45(9):7894–7900. https://doi.org/10.1016/j.matpr.2020.12.615
10. Huu PN, Muthuramalingam T, Van DP, Shirguppikar S, Tien DH, Van TN, Trong LN (2022) Multi-objects optimization in μ-EDM using AlCrNi-coated tungsten carbide electrode for Ti-6AL-4 V. Int J Adv Manuf Technol 122(5–6):2267–2276. https://doi.org/10.1007/s00170-022-10022-8
11. Phan NH, Vu NN, Shirguppikar S, Ly NT, Tam NC, Tai BT, Thanh LTP (2022) Multi-criteria decision making in electrical discharge machining with nickel coated aluminium electrode for titanium alloy using preferential selection index. Manuf Rev 9(13):1–10. https://doi.org/10.1051/mfreview/2022010
12. Gnanasekaran RKK, Raja DE, Jagadeesh S, Singh SP (2022) Experimental study of micro-EDM on EN24 steel with normal brass, tin coated brass, cryogenic treated brass tool by varying the machining parameters. Mater Today: Proc 66(4):2062–2069. https://doi.org/10.1016/j.matpr.2022.05.495
13. Huu PN, Trong LN (2023) Multi-objective optimization in micro-electrical discharge machining using titanium nitride coated WC electrode. Int J Interact Des Manuf 17(8):187–196. https://doi.org/10.1007/s12008-022-01121-7
14. Mandal P, Mondal SC (2019) Development and application of Cu-SWCNT nanocomposite–coated 6061Al electrode for EDM. Int J Adv Manuf Technol 103(7–8):3067–3076. https://doi.org/10.1007/s00170-019-03710-5
15. George J, Chandan R, Manu R, Mathew J (2021) Experimental investigation of silicon powder mixed EDM using graphene and CNT Nano particle coated electrodes. SILICON 13(11):3835–3851. https://doi.org/10.1007/s12633-020-00658-0
16. Mac TB, Shirguppikar S, Ganachari V (2023) The optimization of micro-EDM machining process when using carbon coated micro electrode as a tool. J Mach Eng 23(2):186–197. https://doi.org/10.36897/jme/165929
17. Van DP, Shirguppikar S, Huu PN, Thangaraj M, Phuong TLT, Trong LN, Ganachari V (2023) Influence of graphene coating in electrical discharge machining with an aluminum electrode. J Mech Behav Mater 32:1–10. https://doi.org/10.1515/jmbm-2022-0287

Parametric Analysis of Metal Matrix Syntactic Foams

Imre Fekete and Imre Czinege

1 Introduction

A special group of lightweight materials are syntactic metal foams, in which the reinforcement or filler is usually embedded in a lightweight metal matrix. The accepted name for these materials is "metal matrix syntactic foam", abbreviated as MMSF in the international literature. In principle, the matrix material of MMSFs can be any metal, but in practice light metals such as aluminum and magnesium alloys are mainly used. The filler particles are usually hollow spheres made of ceramic, metal, or glassy materials. Other porous particles such as expanded perlite, clay or glass and fly ash can also be used as fillers. The resulting composite materials have several advantageous properties, in particular, lower density than the matrix material, but also better specific strength, elastic modulus, and good mechanical energy absorption than conventional metal foams.

Research into conventional porous metal foams intensified in the last years of the last century and the first developments reached industrial application at the turn of the millennium. The test results accumulated up to 2000 have been systematically summarized by Ashby et al. in their paper [1] and incorporated into their system of material property charts. In a subsequent publication, they gave a complete survey of the strength, energy absorption, thermal, damping, fatigue, and fracture properties associated with the available metal foams in property charts and derived approximate relationships for their characterization [2]. The period since the 2000s has seen a significant jump in the development of MMSFs. This is illustrated by a few summary publications, such as [3–6]. Afolabia and co-authors [3] present summary tables with references to literature sources, metal and polymer matrix materials, fabrication technologies, and metal foams filled with hollow ceramic spheres and discuss bio- and

I. Fekete · I. Czinege (✉)
Széchenyi István University, Győr, Hungary
e-mail: czinege@sze.hu

G. L. Conte and O. Sename (eds.), *Proceedings of 12th International Conference on Mechatronics and Control Engineering*, Lecture Notes in Mechanical Engineering,
https://doi.org/10.1007/978-981-96-6452-8_2

nanosynthesized foams. A highlight of the paper is the wide range of applications, with a particular focus on the automotive and aerospace industries. The publication [4] also offers a detailed analysis of the manufacturing processes, including a comparison and evaluation of the different casting technologies. It provides information on the properties of fillers and MMSFs and analyzes in detail the matrix–filler interface interactions and reactions. It draws attention on the effect of the strength of the matrix and the spheres to the fracture modes of MMSFs, referring to publications [7] and [8]. Publication [6] illustrates the manufacturing processes of MMSFs and presents the matrix and filler materials in clear charts. It mentions some present studies that analyze the internal structure of MMSFs, highlighting the increasing use of particles of different sizes or materials in addition to the traditional metal foams with a uniform filler. These are referred to in the literature as bimodal or hybrid structures, and the results of a series of experiments are presented in [9].

An important part of the tests suitable for the characterization of the properties of MMSFs is the quasi-static compression test, which was standardized in 2011 [10]. Typical plots of this test are shown in Fig. 1a and b. Figure 1a refers to MMSFs containing hollow metal spheres or other low-strength fillers, while Fig. 1b refers to metal foams filled with brittle ceramic spheres. A comparison of the plots shows that chart in Fig. 1a is related to the plastic behavior of the test specimen, while chart of Fig. 1b shows brittle fracture. The characteristic barreling deformation and fracture along the compression cones occur when the strength of the matrix is dominant relative to the filler. However, when the brittle spheres carry the higher load, the failure shown in Fig. 1b takes the form of a shear fracture at an angle of approximately 45° to the axis [7, 8]. Figure 1b also differs from Fig. 1a in that the initial rise in the former ends in a sharp peak, referred to in the literature as the critical compressive strength (σ_{cr}). The almost constant stress after the peak is the plateau stress (σ_{pl}), which is interpreted as average stress in a given strain range. Such characteristic curve segments cannot be observed in metal foams subject to barrel-like deformation, so the yield stress (σ_y) and the average of the plateau stress are assigned to a given deformation (Fig. 1a). Both figures illustrate the absorbed mechanical energy W, which is proportional to the area under the stress–strain diagram within a chosen strain range. It is also commonly described as the product of the average plateau stress and the strain range.

The failure process described in Fig. 1b is characterized in more detail in a series of papers by Kozma and co-authors [11–13] in which a correlation between the engineering stress–engineering strain chart and the failure of ceramic spheres was determined by analyzing CT images acquired during compressive test.

Several publications contain formulas for the calculation of the formerly presented compressive strength characteristics. Related to this topic, the publications [14–17] and [18] can be mentioned, and will be analyzed in the following, focusing on metal foams filled with hollow ceramic spheres. The relationships proposed by mentioned publications, with uniform notation, are as follows:

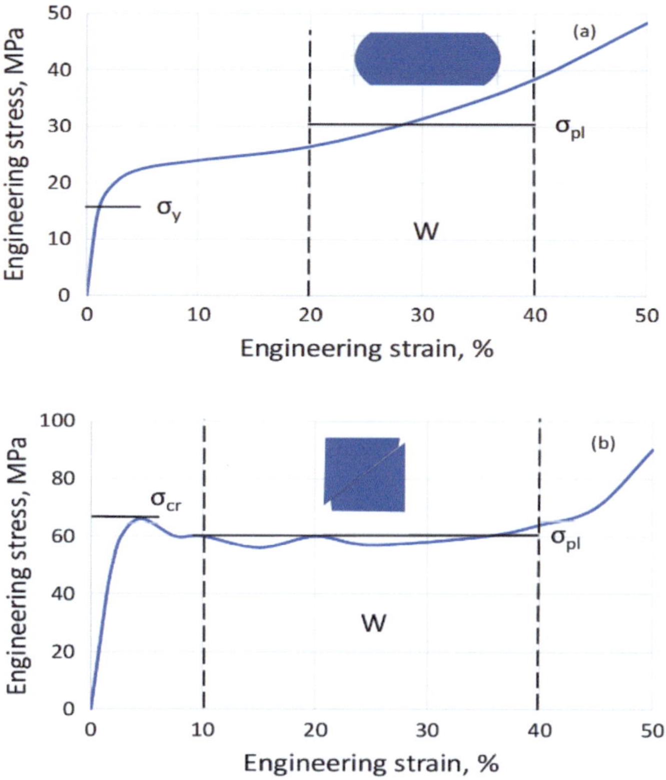

Fig. 1 Compression test charts of syntactic foams. **a** MMSF with metal spheres. **b** MMSF with ceramic spheres

$$[14]\quad \sigma_{cr} = C\sigma_m\left[1 - f_s\left(1 - \frac{2t}{D}\right)^3\right]^n C = 5; n = 3 \tag{1}$$

$$[15]\quad \sigma_{cr} = C\left[\sigma_m f_m^n + \sigma_w f_s\left(1 - \left(1 - \frac{2t}{D}\right)^3\right)^n\right] C = 0,3; n = 3/2 \tag{2}$$

$$[16]\quad \sigma_{cr} = C\left[\sigma_m f_m^n + C\sigma_w\left(\left(1 - \left(1 - \frac{2t}{D}\right)^3\right)(1 - f_{swv})\right)^n\right] C = 0,75; n = 2,19 \tag{3}$$

$$[17]\quad \sigma_{cr} = (2\sigma_m f_m + f_s\eta\sigma_{fw})\left(\frac{A_w}{A_s}\right) \text{ where } \left(\frac{A_w}{A_s}\right) = \left(4\frac{t}{D} - 4\left(\frac{t}{D}\right)^2\right) \tag{4}$$

$$[18]\quad \sigma_{cr=}0,46\sigma_m f_m + \left(1 - \left(1 - \frac{2t}{D}\right)^2\right)(1 - 0,46f_m)\sigma_{Bw} \tag{5}$$

Notations used in the formulas and further discussion:

σ_{cr}: critical compressive strength,

σ_m: yield strength of matrix,

σ_{pl}: plateau stress,

σ_{fw}: theoretical fracture strength of hollow sphere wall,

$\sigma_w = \eta\sigma_{fw}$: real fracture strength of hollow sphere wall,

f_{swv}: volume percentage of voids within the sphere wall,

η: volume percentage of wall material within the sphere wall, $(f_{swv} + \eta = 1)$,

σ_{Bw}: bending strength of hollow sphere wall,

f_m: volume percentage of matrix,

f_s: volume percentage of spheres $(f_m + f_s = 1)$,

D: outside diameter of hollow sphere,

t: wall thickness,

A_w: meridian cross section of hollow sphere wall,

A_s: meridian cross section of sphere,

ρ_{sf}: bulk density of syntactic foam,

C, *n*: material constants,

By analyzing Eqs. (1)–(5), it can be concluded that Eq. (1) does not include the strength of the sphere wall material, but only the wall thickness-to-diameter ratio, and is therefore not very suitable for describing the MMSF. In the derivation of Eqs. (2) and (3), the authors considered the load partitioning within the matrix and the hollow spheres for simple compressive loading, here the third power of the *t*/*D* ratio is included in the formulae. In contrast, the authors of formulae (4) and (5) assumed that the failure is caused by shear stress acting in the approximate 45° plane, where the load is distributed between the active cross section of the matrix in this plane and the meridian cross section of the spherical shells. Therefore, both authors use the A_w/A_s quotient in the equations, which is proportional to the square of *t*/*D*. The presented relations have been validated by the authors with their own experiments, from which the material constants in the formulas have been determined.

It is worth noting that several authors have developed analyses based on the FEM technique, for example, [12] and [19]. Further experimental results and the conclusions drawn from them can be found in the publications listed here [20–27]. From these results, general relationships for MMSFs filled with hollow ceramic spheres can be determined. It can be concluded that the factors that increase the critical compressive strength of MMSF are yield strength of matrix, volume percentage of spheres, fracture strength of hollow sphere wall, t/D wall thickness-to-diameter ratio, while the increase of sphere diameter decreases it. The numerical conclusions on the effect of each factor can be derived mainly from experimental results. Similarly, a number of experimental results are available to characterize the relationships between the output parameters, namely, critical compressive stress, plateau stress, absorbed energy, and densification strain. These show that there is a positive linear relationship between the critical compressive stress and the plateau stress, and that the magnitude of the absorbed energy is increased by all the factors that increased the critical compressive stress. However, there is an inverse proportionality between critical compressive stress and densification strain.

From a brief review of the publications, it can be concluded that the influence of input parameters on the application-relevant properties of MMSFs is known at trend level, but few analyses have addressed the quantified effect of input parameters on output parameters. Therefore, the aim of this paper is twofold, first a parameter sensitivity analysis and then development of process window based on known equations. These results will be used to define the parameter selection strategy.

2 Method of Analysis

2.1 Model Selection

As a first step, the relationships in Eqs. (1)–(5) are analyzed using a database taken from [17]. The data are related to A206-O, A206-T4, A206-T7, and A380-O aluminum alloys as matrix. The mean and min-max values of the input and output data are shown in Table 1. The σ_m matrix yield strength and t/D ratios take discrete values, $\sigma_m = 60, 260$, and 345 MPa for A206 alloy and $\sigma_m = 160$ MPa for A380 alloy. The t/D ratios form two groups, the range 0.052–0.063 is represented later in this paper by the mean 0.06, and an additional $t/D = 0.105$ is included in the database. The values of the f_m matrix volume fraction and σ_{cr} critical compressive strength are randomly located within the respective ranges. The publication contains 61 variants of input parameters and output results. These data have been successfully used by the authors of [17] to verify Eq. (4) and other relationships in the cited paper.

Table 1 Characteristic values of input and output parameters

		σ_m (MPa)	σ_{UTS} (MPa)	D mm	t/D	A_w/A_s	f_m	ρ_{sf} (g/cm^3)
Inputs	Average	205.5	347.1	0.391	0.071	0.264	0.420	1.943
	Minimum	60.0	200.0	0.159	0.052	0.199	0.307	1.770
	Maximum	345.0	435.0	0.925	0.105	0.377	0.548	2.350
		σ_{cr} (MPa)	σ_{min} (MPa)	σ_{pl} (MPa)	ε_t	W/V (J/cm^3)	W/m (J/g)	
Outputs	Average	207.9	134.4	167.8	0.391	61.2	31.5	
	Minimum	144.0	69.0	102.0	0.260	32.7	17.6	
	Maximum	358.0	249.0	278.0	0.530	96.2	45.6	

Using this database, Eqs. (1)–(5) were evaluated. A comparison of the measured values in [17] and those calculated using Eq. (4) is shown in Fig. 2a. This is essentially the same as the diagram reported by the authors, the slope of the line through the origin is close to 1 and the square of the regression coefficient is high (0.9525). However, when the other equations were examined, the calculation with the original material parameters showed significant deviations from the measured values, as shown in the lower data plot of Fig. 2b for Eq. (2) [15]. Here the slope of the line was only 0.3 with the original material parameter C = 0.3. However, the approximate value of C was found by minimizing the sum of squared error between the measured and calculated values (C = 1.007), so that the slope of line approached 1.

By performing similar calculations for the other equations, the appropriate material parameters could be determined. The original and modified parameters and the goodness of fits for Eqs. (2)–(5) are given in Table 2. The calculation with Eq. (1) is not included here because it yielded very low correlation coefficients for all parameter variants. However, the other equations, when tested on the same dataset after adjusting the material parameters, gave a reasonable approximation. We should like

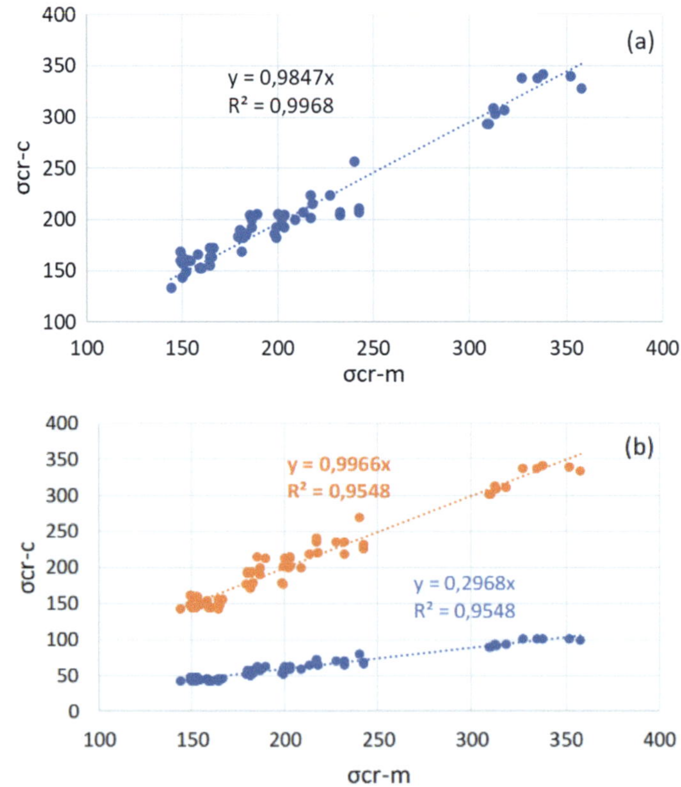

Fig. 2 Comparison of measured and calculated values. **a** Eq. (4). **b** Eq. (2)

Table 2 Comparison of equations

No. of equations	Slope	R^2	Original	Applied
(2)	0.997	0.955	$C = 0.3$; $\sigma_w = 1154$ MPa	$C = 1.007$; $\sigma_w = 1154$ MPa
(3)	0.997	0.911	$C = 0.75$; $n = 2.19$; $\sigma_w = 2944$ MPa; $V_{swv} = 0.5$	$C = 0.75$; $n = 1.384$; $\sigma_w =$ 2944 MPa; $V_{swv} = 0.5$
(4)	0.985	0.953	$\eta^*\sigma_w = 1051$ MPa	$\eta^*\sigma_w = 1051$ MPa
(5)	0.992	0.873	$\sigma B_{.w}$= 205 MPa	$\sigma_{B.w} = 800.7$ MPa

to emphasize that the objective of this parameter correction is not the criticism of the equations, as they were elaborated using different test materials and testing methods, only to harmonize the equations on a common database. This is necessary only for the correct evaluation of sensitivity analysis.

2.2 Methods for Analyzing the Impact of Parameters

If known relationships, supported by experimental results, are available between the properties of the components building MMSF and the output characteristics determined by the compression test, then important conclusions can be drawn from the relationships themselves. This can be done using parameter sensitivity analysis, which is a useful tool for exploring the relationships between different independent variables (inputs) and the dependent variable (output). The simplest formula is to express the sensitivity of an output variable to the input parameter under study: *sensitivity = percentage change in output/percentage change in input*. Thus, when conducting a sensitivity analysis, one must first identify the relationship between the key variables affecting the output, and then determine how the output changes as a result of changes in the inputs. In this paper, Eqs. (2)–(5) and the Response Surface Methodology (RSM) approximation are used to perform the analysis. RSM is a regression model that fits a polynomial function to the measurement data. In our own calculations, we used a quadratic polynomial approach. There is a rich literature on this method and several software provide ready-made solutions for the calculations, so we refer to only two publications [28, 29] instead of a detailed description.

It has been verified previously that each formula correctly describes the relationship between the four input parameters (σ_m. σ_w, f_m, t/D) and the output (σ_{cr}), so that the relationships can be used to evaluate the sensitivity. The RSM technique also gives a function that approximates the measurement results well ($R^2 = 0.931$), so the use of this technique provides an additional comparison. A further analysis can be carried out to investigate the influence of the matrix and the filler properties on the critical compressive strength, this is done by individually evaluating the two terms in Eq. (4) and quantifying the combined effect of the two factors in the next section.

2.3 Concept and Application of the Process Window

In general, the process window describes the range of values within which the input variables of a process can be located so that the process could be successfully carried out. When new alloys or composites are developed, there are also constraints, for example, solubility limits and rules on the mixing ratios of certain components. These limiting conditions are visualized as the relationship between two input variables, where the boundary values of the parameters are represented by a surface in an x-y-coordinate system. In the case of more than two variables, it is usual to extend the two-dimensional approximation with parametric variables.

Similarly, the input parameters of MMSFs characterized by Eqs. (2)–(5) cannot be taken arbitrarily, but only under certain restrictive conditions. Obviously, there is a lower limit to the matrix yield strength that can still hold the filling spheres together. For aluminum matrix, this is approximately $\sigma_m = 30$ MPa for the Al99.5 alloy, while the alloys used in other applications can reach an upper limit of 500 MPa in the heat-treated versions of aluminum alloys (T4, T7). The experimentally proven matrix yield strength range of magnesium alloys is 150–350 MPa. Zn alloys are reported in the range 500–700 MPa. The theoretical limit for filler sphere ratio is $f_s = 0.74$ for the hexagonal arrangement with the densest fit and the known $f_s = 0.63$ ratio is generally accepted for the random arrangement. A minimum filler level of $f_s = 0.5$ is typical. For the hollow ceramic sphere wall material, the databases give compressive strength values around 3000 MPa, which decreases to ~1000 MPa when the wall porosity is taken into account, as shown in [17], but a similar value is also given in [15]. Finally, there are data for the *t*/*D* wall thickness ratio from 0.03 to 0.15, but higher ratios can be found in publications such as [14]. All these limits are extreme values, but within these limits there are also constraints on the input parameters.

The boundary conditions between the four input variables previously defined for MMSFs are generally characterized by the range of validity of Eqs. (1)–(5). A good example of this is publication [17], where the authors give the ratio of meridian cross section of sphere wall to meridian cross section of sphere in Eq. (6), the critical cross-sectional ratio $(A_w/A_s)^*$ defines the range of validity of Eq. (4).

$$\left(\frac{A_w}{A_s}\right)^* = \frac{f_m\sigma_m}{2f_m\sigma_m + f_s\eta\sigma_{fw}}\ Eq(4)\ is\ valid\ if \left(\frac{A_w}{A_s}\right)^* \leq \left(\frac{A_w}{A_s}\right) \tag{6}$$

This consideration is based on the assumption that the critical compressive stress of an MMSF is determined by the ratio and strength of the filling spheres, rather than by the load-carrying capacity of the matrix.

3 Results

3.1 Parameter Sensitivity Testing

Parameter sensitivity indices are calculated using formulas (2)–(5) and the RSM technique. Each equation contains the four input parameters described earlier and the output is the critical compressive stress. In fitting the quadratic polynomial according to RSM, three of the input parameters were taken into account, these are the matrix yield strength σ_m, the wall thickness ratio *t*/*D,* and the volume fraction f_s of the spheres, while the output parameter was the critical compressive stress σ_{cr} here also. In the analysis, the average of the input parameters as shown in Table 1 is taken as the initial value and varied by ± 10 and $\pm 20\%$ for each parameter separately. Using these values, the corresponding outputs are determined. Doing this for each parameter gives the diagram shown in Fig. 3a calculated from Eq. (4). Although Eqs. (2)–(5) are not linear for *t*/*D*, they can be considered approximately linear over the range under consideration because *t*/*D* is relatively small ($t/D < 0.15$). From these data, an additional characteristic can be determined, the sensitivity defined earlier, which is the ratio of the variation of each output and input parameter. These parameters are shown in Fig. 3b, determined from Eqs. (2) to (5), and the three values calculated from the RSM approximation are also included in bar chart.

From the results, it can be concluded that the *t*/*D* ratio and strength of the sphere wall ($\eta * \sigma_{fw}$) are the determining factors in the evolution of the compressive strength, the matrix strength (σ_m) has less effect on the output, while the volume ratio of spheres f_s shows significant differences among the equations. The sensitivity to f_s calculated from Eqs. (3) and (4) is above 0.3, whereas from Eqs. (2) and (5) it follows that this parameter has a significantly smaller effect on the output. The RSM approximation results for the *t*/*D* ratio and matrix yield strength agree well with the values obtained from Eqs. (2) to (5), and the effect of f_s is closer to Eqs. (2) and (5). However, it is still clear that the largest scatter is in the estimation of this parameter.

In order to further analyze the uncertainty outlined above, Principal Component Analysis (PCA) was applied to the four parameters already modeled by the RSM analysis. This technique is a well-known multivariate analysis which is used to reduce the dimensionality of data. As a result of PCA, the parameter's characteristic dimensions can be reduced significantly while retaining most of the original data variance. The method uses a suitable transformation to form principal components from the covariance matrix of the measurement results using its eigenvalues (λ_i) and eigenvectors (v_i). The publication [30] contains an example of a general description of the method, and [31] shows a possible application of PCA to the analysis of machining parameters of titanium alloys.

Figure 4a shows the λ_i eigenvalues normalized to 1 and their cumulative values, showing that the first two principal components contain information in the four parameters at a ratio of 0.814, making the two principal components PC1 and PC2 suitable for further analysis. Figure 4b illustrates the coefficients of the first two

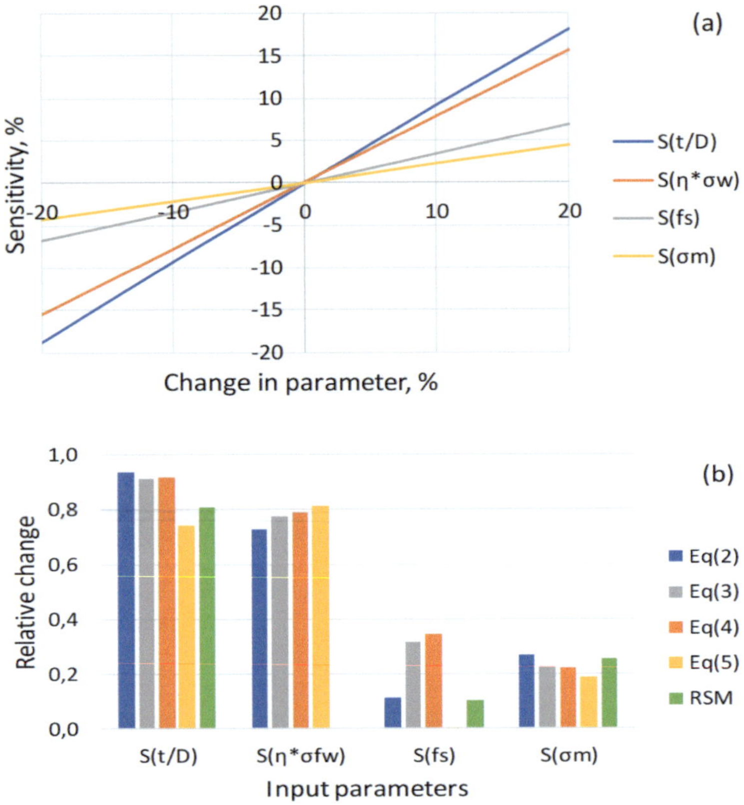

Fig. 3 Parameter sensitivity charts. **a** Characteristic sensitivity values for Eq. (4). **b** Comparison of different calculations

eigenvectors, from which it can be seen, that the coefficients of the critical compressive strength and wall thickness ratio are the largest for PC1, while the yield strength of the matrix is the dominant one in PC2 (Fig. 5).

3.2 Effect of Matrix and Filler on the Critical Compressive Stress

In the following, the load-bearing capacity of the two components of Eq. (4) is analyzed. By splitting Eq. (4) into two parts and using the relationship $f_s = 1 - f_m$, the contribution of the matrix and filler to the critical compressive stress is as follows:

$$\sigma_{m-cr} = 2\sigma_m(1 - f_s)\left(\frac{A_w}{A_s}\right) \text{ and } \sigma_{w-cr} = f_s \eta \sigma_{fw}\left(\frac{A_w}{A_s}\right) \quad (7)$$

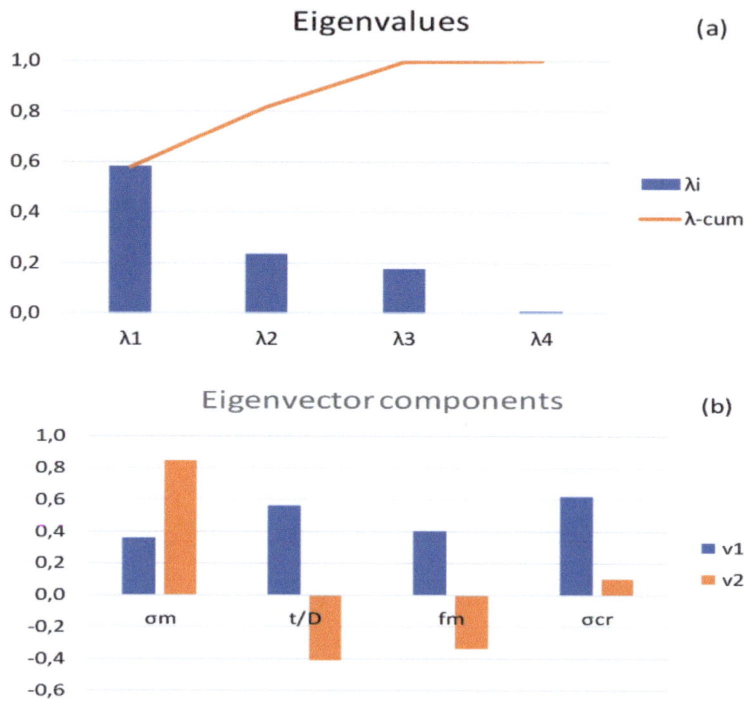

Fig. 4 Eigenvalues and eigenvectors

Both parts of Eq. (7) represent a straight line as a function of f_s. If the slope of the line $\sigma_{m\text{-}cr}$ is smaller in absolute value than that of $\sigma_{w\text{-}cr}$, then σ_{cr} tends to increase as a function of f_s, as the amount of filler spheres increases, which increases the strength of the MMSF. On the other hand, if the opposite is true, then the matrix strength effect will dominate, and the critical compressive stress will decrease as f_s increases. If the two slopes are equal, σ_{cr} does not vary as a function of f_s. This condition is satisfied when $2\sigma_m = \sigma_w = \eta * \sigma_{fw}$. This is the mathematical formulation of the statement known from the literature under which condition the role of the matrix or filler is decisive in the determination of the properties of the MMSF. These relations are illustrated in Fig. 6, where, using the values $\eta = 0.357$ and $\sigma_{fw} = 2945$ MPa reported in [17] the calculated sphere wall strength is $\sigma_w = \eta * \sigma_{fw} = 0.357 \times 2945 = 1051$ MPa and the matrix strength is $\sigma_m = 420$ MPa in Fig. 6a and $\sigma_m = 700$ MPa in Fig. 6b. Figure 6a shows the contribution of the two parts of Eq. (7) to the compressive strength of the MMSF for the case $2\sigma_m < \sigma_w$ ($2*420 < 1051$) and Fig. 6b for the case $2\sigma_m > \sigma_w$ ($2*700 > 1051$).

A similar conclusion can be reached by analyzing the role of the matrix and the spheres in the load bearing as a function of the ratio $c = \sigma_w/\sigma_m$. To investigate this, both sides of Eq. (4) are divided by σ_m to obtain the critical compressive strength/matrix yield strength ratio on the left and the σ_w/σ_m ratio in Eq. (8) on the right.

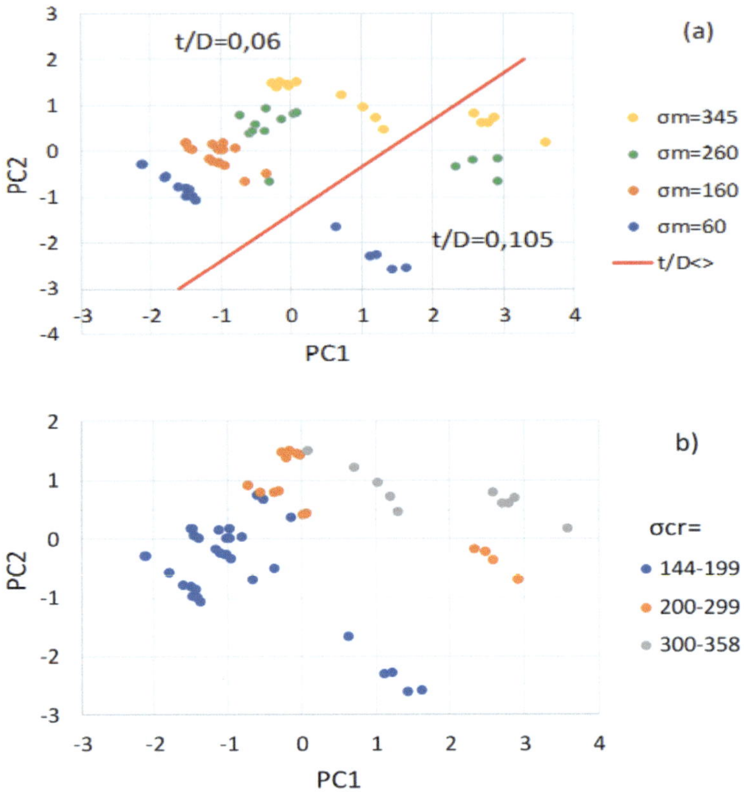

Fig. 5 Analysis of principal components. **a** Effect of matrix yield strength and sphere wall thickness ratio. **b** Arrangement of critical compressive strength values

Plotting the contribution of the matrix and filler spheres to the σ_{cr}/σ_m gives the plots shown in Fig. 7.

$$\frac{\sigma_{cr}}{\sigma_m} = \left(2f_m + f_s\frac{\sigma_w}{\sigma_m}\right)\left(\frac{A_w}{A_s}\right) \tag{8}$$

Figure 7a shows the σ_{cr}/σ_m values according to Eq. (8) as a function of σ_w/σ_m taking into account the parameters seen in the figure, namely, the matrix volume fraction, the wall strength of filling spheres, and the t/D ratio. It can be seen that according to Eq. (8), the contribution of the matrix is constant and with an increase in the sphere wall strength/matrix strength ratio, the contribution of the spheres increases linearly. Figure 7b shows the relative contribution derived from the previous figure, divided by the resulting critical compressive strength values from Fig. 7a. This provides a better illustration of the load-bearing relationship between matrix and filler. It can also be seen from Fig. 7b that the 50–50% load-bearing ratio occurs at $\sigma_w/\sigma_m = 1.175$. This is a function of the volume fraction of the matrix (in this case $f_m = 0.37$),

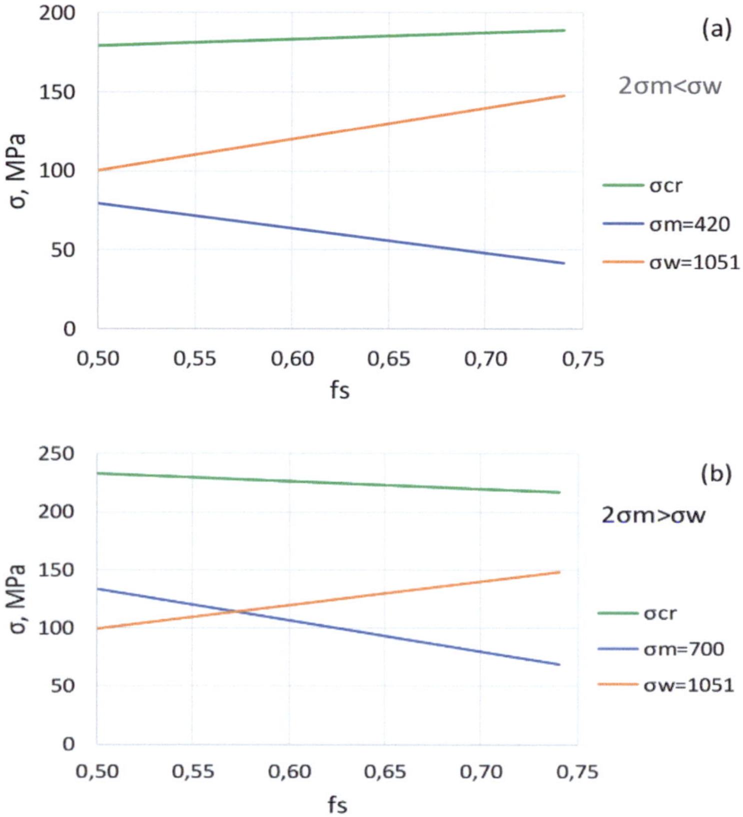

Fig. 6 Load partitioning of matrix and filler material to compressive strength. **a** Effect of spheres is stronger. **b** Effect of matrix is dominant

but if $f_m = 0.5$ then $\sigma_w/\sigma_m = 2$. Note that the initial value $\sigma_w/\sigma_m = 0.3$ shown in the figure can only be obtained for minimum wall strengths, for example, glass spheres. For ceramic spheres, the ratio $\sigma_w/\sigma_m > 2$ is typical. In this range, the load-bearing capacity of the filler spheres is already dominant, as confirmed by the parameter sensitivity study. In this context, it is worth mentioning that the parameter sensitivity analysis showed an approximate ratio of 1:4 between the effect of the filling sphere properties and the matrix strength, and the same can be seen in Fig. 7b for the σ_w/σ_m ratio above 4.

3.3 Definition of the Process Window

The methodology section describes the interpretation of the process window and the constraint given in Eq. (4) as a condition for Eq. (6). This is analyzed in detail

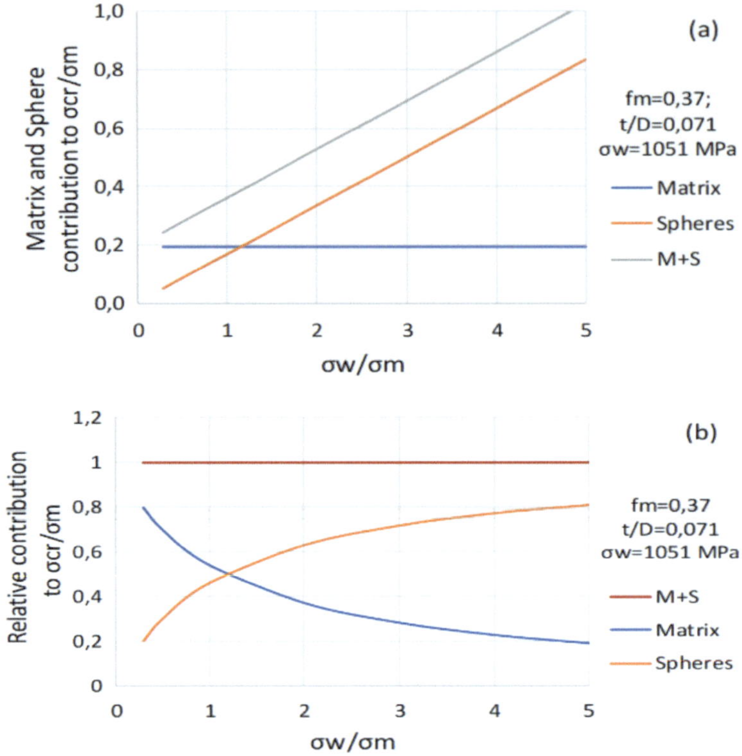

Fig. 7 Effect of sphere wall strength/matrix strength coefficient on the relative compressive strength. **a** Absolute values. **b** Relative values

below. Equation (6) contains four influencing factors. By using the relationship $f_s = 1 - f_m$ and expressing σ_w using the previously introduced "*c*" ratio ($\sigma_w = c*\sigma_m$) and substituting them into Eqs. (6, 9), is obtained, which now contains only two variables:

$$\left(\frac{A_w}{A_s}\right)^* = \frac{f_m}{f_m(2-c)+c} \quad (9)$$

Since it is more convenient to express the geometric characteristic of the filling sphere in terms of the *t*/*D* ratio rather than the surface area ratio, the following equation can be used for the conversion:

$$\left(\frac{t}{D}\right)^* = \frac{1}{2}\left(1 - \sqrt{1 - \left(\frac{A_w}{A_s}\right)^*}\right) \quad (10)$$

Substituting the coefficient $(A_w/A_s)^*$ defined by Eq. (9) into Eq. (10) gives the function $(t/D)^* = f(c, f_m)$, which can be represented by the surface in Fig. 8a or in parametric form in Fig. 8b. The process window in 3D shown in Fig. 8a can be represented such that any value of t/D above the surface defined by the two parameters can be taken as the allowed value. The same can be deduced from Fig. 8b for the three characteristic matrix volume ratios, where the values above each curve satisfy the condition $(A_w/A_s)^*$ in Eq. (6).

Equation (6) can also be analyzed from a different point of view, because if we express the matrix strength σ_m from the equation, we obtain the minimum matrix yield strength ($\sigma_{m\text{-min}}$) that satisfies $(A_w/A_s)^*$ for a given value of f_m and σ_w. This relationship is shown in Eq. (11):

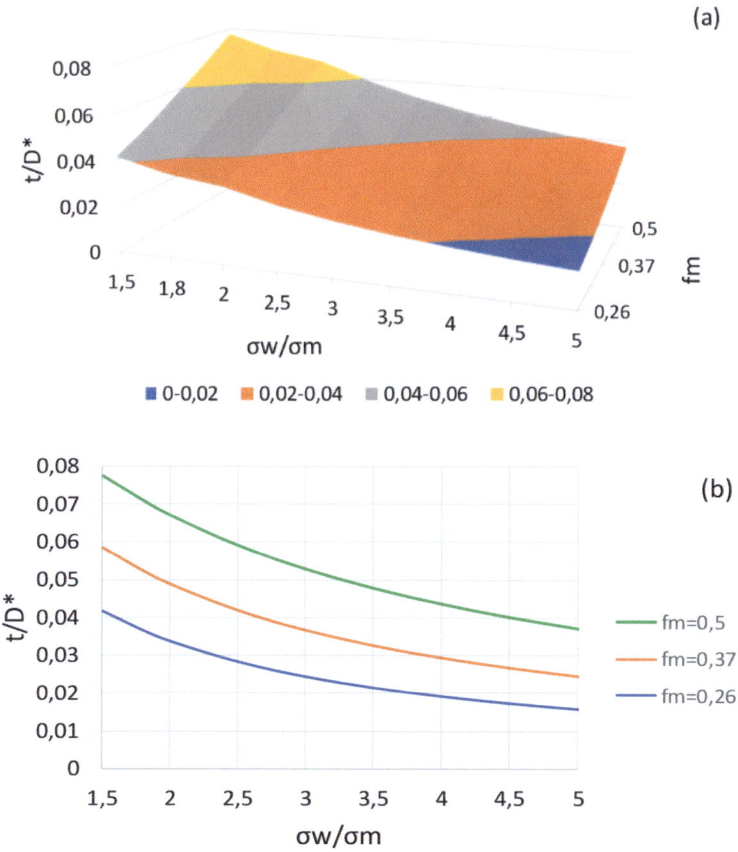

Fig. 8 Process window for thickness/diameter ratio of sphere. **a** 3D illustration. **b** Parametric functions

$$\sigma_{m-\min} = \frac{\left(\frac{A_w}{A_s}\right)^*(1-f_m)\eta\sigma_{fw}}{f_m\left(1-2\left(\frac{A_w}{A_s}\right)^*\right)} \tag{11}$$

Knowing this minimum matrix strength, a new process window can be constructed, which now contains the critical compressive stress value as a function of the matrix strength and the ratio of the filling sphere wall-to-diameter ratio, as shown in Fig. 9a. The parameters included here are the matrix volume fraction and the compressive strength of the sphere wall material. It can be seen from the figure that the three smallest *t*/*D* values (0.02, 0.04, 0.06) in the specific matrix strength range are affected by the $\sigma_{m\text{-min}}$ condition, which limits the applicable matrix strength. The *t*/*D* ratios above 0.08 allow the full range of matrix strength to be used.

Evaluating the effect of all the influencing factors together, it can be concluded that the volume fraction of the spheres, the strength of the sphere wall, the wall thickness

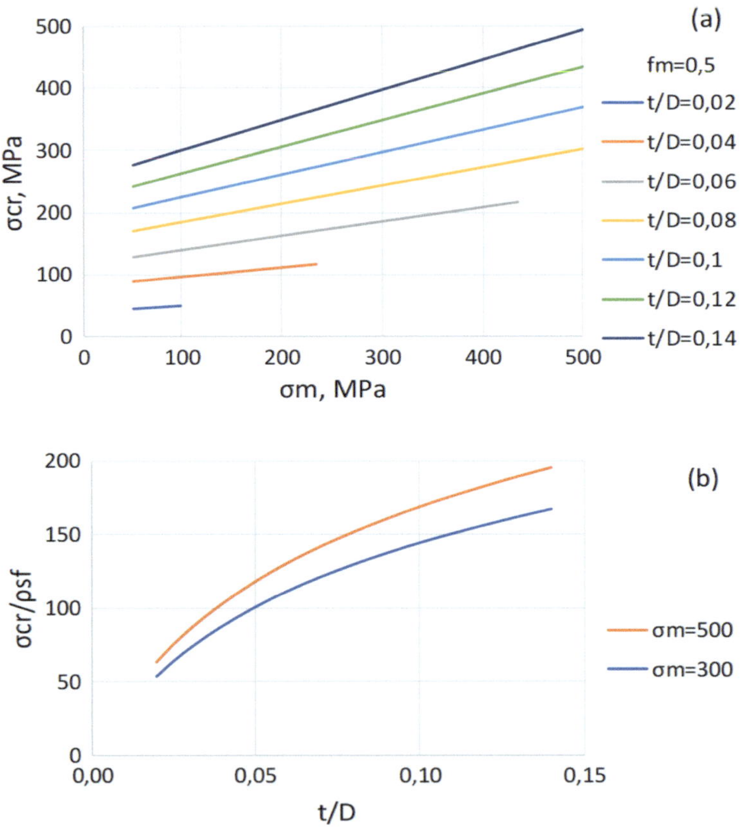

Fig. 9 Charts for selecting optimal compression strength. **a** Limits of matrix strength. **b** Relative compressive strength values

ratio, and the strength of matrix all increase the critical compressive strength of the MMSF. A limiting factor may be whether an increase in the volume fraction and wall thickness ratio of the filling spheres will not excessively increase the density of the MMSF, as this will lead to an increase in the mass of the structure. This is answered in Fig. 9b, which shows the specific compressive strength (σ_{cr}/ρ_{sf}) as a function of the t/D ratio, with two matrix strengths as parameters. From the calculations and the graph, it can be seen that as the density increases by a factor of two, the critical compressive strength increases by a factor of six, thus the specific strength also increases by a factor of three, so that an increase in the t/D ratio is more effective in promoting strength growth than mass growth.

The last diagrams presented can be used to determine the desired parameters of the MMSF as required. For example, if a given critical compressive strength value is required in a structure, then the t/D-σ_m pairs can be selected by picking them in Fig. 9a. For other boundary conditions, the charts in Fig. 8 can also be used for preliminary parameter selection.

4 Conclusions

In our research, we have analyzed the equations for the calculation of the strength properties of MMSFs determined by compression test and evaluating them on a common database. It was found that, with the appropriate choice of material parameters, the selected equations are in good agreement with the experimental results and are therefore suitable for further analysis. Based on these analyses, the following conclusions were reached:

A parameter sensitivity analysis was carried out to analyze the effect of each factor on the critical compressive strength and it was found that the ratio of the sphere wall thickness to diameter and the wall strength have approximately four times greater effect on the critical compressive strength than the yield strength of the matrix. Uncertainty in the estimation of the matrix volume fraction is indicated by the results obtained from the equations studied and by the RSM approximation.

The principal component analysis demonstrated that the first two principal components contain more than 80% the compressed information of the four parameters. The matrix strength and the sphere wall thickness-to-diameter ratio show the highest selectivity in the principal component coordinate system. Similarly, the critical compressive strength values were also found to be clustered.

By analyzing the load-bearing capacity of the matrix and the filler spheres, it has been shown that for MMSFs, the load-bearing ratio of the hollow ceramic spheres is dominant when the strength ratio of the sphere wall to the matrix is greater than 2. This strength ratio is most favorable for the critical compressive strength at values between 4 and 5. The statement laid down here corresponds well with a similar measure of parameter sensitivity.

For the parameter selection of MMSFs, process windows were constructed to show the minimum wall thickness/diameter ratio as a function of the strength ratio of the spherical wall to matrix and the volume ratio of the matrix, and to determine the minimum matrix strength.

It was found that an increase in the wall thickness/diameter ratio, which greatly increases the critical compressive stress, does not adversely increase the density of the MMSF, and the specific critical compressive strength also increases markedly with an increase in the wall thickness/diameter ratio.

Acknowledgement The research presented in this paper was funded by the National Security Subprogram at the Széchenyi István University (TKP2021-NVA-23).

References

1. Ashby MF, Evans AG, Fleck NA, Gibson LJ, Hutchinson JW, Wadley HNG (2000) Metal foams: a design guide. Butterworth-Heinemann
2. Ashby MF (2011) Materials selection in mechanical design. Elsevier Butterworth Heinemann. Fourth edition
3. Afolabia LO, Ariff ZM, Saiyid Hashim SF, Alomayri T, Mahzana S, Kamarudina K-A, Muhammad ID (2020) Syntactic foams formulations, production techniques, and industry applications: a review. J Mater Res Technol 9(5):10698–10718
4. Bolat Ç, Akgün İC, Gokşenli A (2020) On the way to real applications: aluminum matrix syntactic foams. Eur Mech Sci 4(3):131–141
5. Duarte I, Fiedler T, Krstulovic-Opara L, Vesenjak M (2020) Brief review on experimental and computational techniques for characterization of cellular metals. Metals 10:726
6. Orbulov IN (2021) Metal matrix composite syntactic foams for light-weight structural materials. In: Reference module in materials science and materials engineering
7. Balch DK, O'Dwyer JG, Davis GR, Cady CM, Gray III GT, Dunand DC (2005) Plasticity and damage in aluminum syntactic foams deformed under dynamic and quasi-static conditions. Mater Sci Eng A 391:408–417
8. Balch DK, Dunand DC (2006) Load partitioning in aluminum syntactic foams containing ceramic microspheres. Acta Materialia 54(6):1501–1511
9. Kemény A, Leveles B, Bubonyi T, Orbulov IN (2020) Effect of particle size and volume ratio of ceramic hollow spheres on the mechanical properties of bimodal composite metal foams. Compos Part A-Appl Sci Manuf 140:106152
10. ISO13314:2011 Mechanical testing of metals—Ductility testing—Compression test for porous and cellular metals
11. Kozma I, Fekete, Zsoldos I (2017) Failure analysis of aluminium—ceramic composites. Mater Sci Forum 885:286–291
12. Kozma I, Zsoldos I (2019) CT-based tests and finite element simulation for failure analysis of syntactic foams. Eng Fail Anal 104:371–378
13. Kozma I, Harangozó D, Hargitai H, Zsoldos I (2019) Displacement based failure analysis of metal matrix syntactic foam. WIT Trans Eng Sci
14. Kiser M, He MY, Zok FW (1999) The mechanical response of ceramic microballoon reinforced aluminum matrix composites under compressive loading. Acta mater 47(9):2685 ± 2694
15. Wu GH, Dou ZY, Sun DL, Jiang LT, Ding BS, He BF (2007) Compression behaviors of cenosphere–pure aluminum syntactic foams. Scr Mater 56:221–224
16. Mondal DP, Ramakrishnan DS, Uday N, Bhasker K (2009) Cenosphere filled aluminum syntactic foam made through stir-casting technique. Compos Part A 40:279–288
17. Ferguson JB, Santa Maria JA, Schultz BF, Rohatgi PK (2013) Al–Al2O3 syntactic foams—part II: predicting mechanical properties of metal matrix syntactic foams reinforced with ceramic spheres. Mater Sci Eng: A

18. Hartmann M (2015) Herstellung, Struktur und Eigenschaften syntaktischer Magnesiumschäume. PhD Thesis. Der Technischen Fakultät der Friedrich-Alexander-Universität Erlangen-Nürnberg
19. Li P, Petrinic N, Siviour CR (2012) Finite element modelling of the mechanism of deformation and failure in metallic thin-walled hollow spheres under dynamic compression. Mech Mater 54:43–54
20. Alejandro G, Rivero R, Schultz BF, Ferguson JB (2013) Compressive properties of Al-A206/SiC and Mg-AZ91/SiC syntactic foams. J Mater Res. https://doi.org/10.1557/jmr.2013.176
21. Luong DD, Strbik III OM, Hammond VH, Gupta N, Cho K (2013) Development of high performance lightweight aluminum alloy/SiC hollow sphere syntactic foams and compressive characterization at quasi-static and high strain rates. J Alloy Compd 550:412–422
22. Santa Maria JA, Schultz BF, Ferguson JB, Rohatgi PK (2013) Al-Al_2O_3 syntactic foams—part I.: effect of matrix strength and hollow sphere size on the quasi-static properties of Al-A206/Al_2O_3 syntactic foams. Mater Sci Eng A 582:415–422
23. Santa Maria JA, Schultz BF, Ferguson JB, Gupta N, Rohatgi PK (2014) Effect of hollow sphere size and size distribution on the quasi-static and high strain rate compressive properties of Al-A380–Al_2O_3 syntactic foams. J Mater Sci 49:1267–1278. https://doi.org/10.1007/s10853-013-7810-y
24. Orbulov IN, Szlancsik A (2018) On the mechanical properties of aluminum matrix syntactic foams. Adv Eng Mater 20(5)
25. Movahedi N, Murch GE, Belova IV, Fiedler T (2019) Functionally graded metal syntactic foam: fabrication and mechanical properties. Mater Des 168:107652
26. Shilkoa SV, Chernous DA, Zhang Q, Lin Y, Choe H (2020) Uniaxial compression model for a metal-matrix/hollow-microsphere composite synthesized by pressure infiltration. Mech Mater 144:103349
27. Su M, Wang H, Hao H, Fiedler T (2020) Compressive properties of expanded glass and alumina hollow spheres hybrid reinforced aluminum matrix syntactic foams. J Alloy Compd 821:153233
28. Hadiyat MA, Sopha BM, Wibowo BS (2022) Response surface methodology using observational data: a systematic literature review. Appl Sci 12:10663. https://doi.org/10.3390/app122010663
29. Marques AE, Prates PA, Pereira AFG, Oliveira MC, Fernandes JV, Ribeiro BM (2020) Performance comparison of parametric and non-parametric regression models for uncertainty analysis of sheet metal forming processes. Metals 10:457. https://doi.org/10.3390/met10040457
30. Marukatat S (2022) Tutorial on PCA and approximate PCA and approximate kernel PCA. Artif Intell Rev. https://doi.org/10.1007/s10462-022-10297-z
31. Dominguez LA, Shokrani A, Flynn JM, Dhokia V, Newman ST (2019) Application of multivariate statistical analysis for CNC milling of large Ti-6Al-4V components. Procedia Manuf 38:800–807

Computational Analysis for Failure of a Metal Rod

Suchakree Euaumpon, Vichayuth Satja, and Thira Jearsiripongkul

1 Introduction

The standard inspection method for a connecting rod is scoping. However, it is relatively expensive and time-consuming. This makes it difficult for companies to conduct frequent inspections of their equipment. In addition, internal cracks within the rod are hidden inside the material [1, 2]. Thus, it cannot be found by a standard scoping method. Mogal Shyam Prabhakar [3] analyzed the vibration of cracked beams. He suggested that the reduction of the system's natural frequency along with its vibrational mode shape can be used to identify possible locations of hidden cracks.

Subramani et al. [4] investigated crack identification in reinforced concrete beams. Studies have shown that beams, made of brittle materials like concrete or cement, show increasing crack development during their service life due to mechanical and environmental loadings [5].

Kunla et al. [6] studied the identification of the crack location in the biomaterial metal cantilever beam under dynamic load based on the center difference approximation.

Identification of open crack of beam using model based method was carried out by Li et al. [7]. They proposed the "Model-Based Method" method to identify the depth and spacing of cracks, which is capable of estimating the general displacement of all nodes on the beam under a restricted degree of freedom.

In this study, we propose the use of vibrations to detect the presence of cracks is suggested. This method is also capable of identifying internal cracks which are known to be hidden from previous methods of inspection. A new testing method may help shorten the time required for an inspection at a cheaper cost.

S. Euaumpon (✉) · V. Satja · T. Jearsiripongkul
Mechanical Engineering Department, Thammasat University Pathum Thani, Pathum Thani, Thailand
e-mail: suchakree.eua@dome.tu.ac.th

G. L. Conte and O. Sename (eds.), *Proceedings of 12th International Conference on Mechatronics and Control Engineering*, Lecture Notes in Mechanical Engineering,
https://doi.org/10.1007/978-981-96-6452-8_3

2 Analysis

The first step to derive the eigenfunction required to determine the natural frequency of a cracked fixed-free rod undergoing a free vibration is to determine the governing equation of the longitudinal vibration of a nanobeam. Using nonlocal elastic theory [8]. The governing equation of the longitudinal vibration of a nanobeam could then be written as:

$$EA\frac{\partial^2 U}{\partial X^2} = m\frac{\partial^2 U}{\partial t^2}\left[1 - (e_0 a)^2 \frac{\partial^2}{\partial X^2}\right] \tag{1}$$

where m is the mass per length, U is the longitudinal displacement, $e_0 a$ represents the non-local constant, t represents time, and E and A represent the material's modulus of elasticity and cross-sectional area respectively. A model of longitudinal vibration of cracked nanobeams was considered using nonlocal elastic theory [8], as shown in Fig. 1.

The cracked could be considered as 2 separate rods therefore governing equation could then be written as follows:

$$EA\frac{\partial^2 U_1}{\partial X^2} = m\frac{\partial^2 U_1}{\partial t^2}\left[1 - (e_0 a)^2 \frac{\partial^2}{\partial X^2}\right] \quad 0 \le X < C \tag{2}$$

$$EA\frac{\partial^2 U_2}{\partial X^2} = m\frac{\partial^2 U_2}{\partial t^2}\left[1 - (e_0 a)^2 \frac{\partial^2}{\partial X^2}\right] \quad C < X \le L \tag{3}$$

There are a total of 4 boundary conditions required to solve for the natural frequency of a fixed-free cracked rod undergoing a free vibration as follows:

$$U_1(0,t) = 0, \quad k\left[U_1(C,t) - U_2(C,t)\right] = -\left[EA + (e_0 a)^2 m \frac{\partial^2}{\partial t^2}\right]\frac{\partial U_1}{\partial X},$$
$$\frac{\partial U_1(C,t)}{\partial X} = \frac{\partial U_2(C,t)}{\partial X} \quad and \quad U_2^{'}(L,t) = 0$$

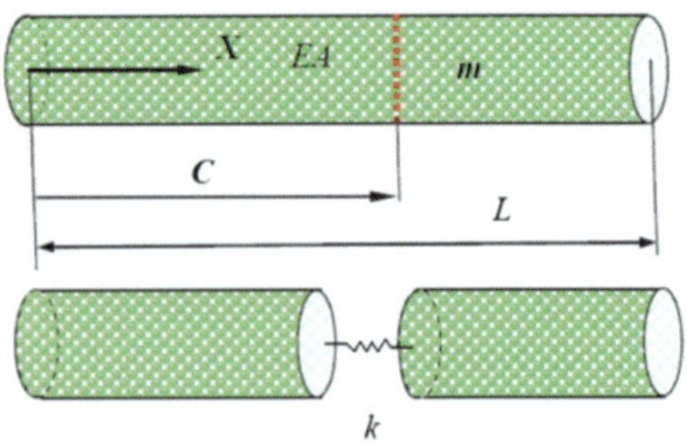

Fig. 1 Model of the cracked rod [8]

By applying the separation variable method we obtained:

$$\frac{\partial^2 W_1(X)}{\partial X^2}\left[EA - m\omega^2(e_0a)^2\right] = -mW_1(X)\omega^2 \quad 0 \le X < C \tag{4}$$

$$\frac{\partial^2 W_2(X)}{\partial X^2}\left[EA - m\omega^2(e_0a)^2\right] = -mW_2(X)\omega^2 \quad C < X \le L \tag{5}$$

To convert the equations and boundary conditions above into the dimensionless form the following dimensionless constants are assumed.

$$x = \frac{X}{L}, u_1 = \frac{W_1}{L}, u_2 = \frac{W_2}{L}, \epsilon = \frac{\varepsilon_0 a}{L}, b = \frac{C}{L}, \beta = \sqrt{\frac{\omega^2 mL^2}{EA}} \text{ and } K = \frac{EA}{kL}$$

The dimensionless form of Eqs. 4 and 5 could then be written as:

$$\frac{d^2u_1}{dx^2} - \frac{\beta^2}{1-\beta^2\varepsilon^2}u_1 = 0 \quad 0 \le x < b \tag{6}$$

$$\frac{d^2u_2}{dx^2} - \frac{\beta^2}{1-\beta^2\varepsilon^2}u_{22} = 0 \quad b < x \le 1 \tag{7}$$

The boundary conditions could then be converted into dimensionless form as follows:

$$u_1(0) = 0, \quad u_1(b) + K\left[1 - \epsilon^2\beta^2\right]u_1^{'}(b) = u_2(b),$$
$$u_1^{'}(b) = u_2^{'}(b) \quad and \quad u_2^{'}(1) = 0$$

By applying the separation variable method, the general solution of both parts of the cracked rod is as follows:

$$u_1(x) = A_1 cos(\Omega x) + B_1 sin(\Omega x) \tag{8}$$

$$u_2(x) = A_2 cos(\Omega x) + B_2 sin(\Omega x) \tag{9}$$

where $\Omega = \sqrt{\frac{\beta^2}{1-\epsilon^2\beta^2}}$.

The eigenfunction for the longitudinal vibration of a cracked fixed-free rod undergoing a free vibration is as follows:

$$cos\left(\frac{\beta}{\sqrt{1-\epsilon^2\beta^2}}\right) - K\beta\sqrt{1-\epsilon^2\beta^2}cos\left(\frac{b\beta}{\sqrt{1-\epsilon^2\beta^2}}\right)sin\left(\frac{\beta - b\beta}{\sqrt{1-\epsilon^2\beta^2}}\right) \tag{10}$$

Assume that $\epsilon = 0$ and $K = 1$. By plotting the graph of Eq. 10 collect the positions where the curve intercept with the axis then substitute the positions into the following equation:

$$\omega = \sqrt{\frac{\beta^2 EA}{L^2 m}}. \tag{11}$$

3 Results and Discussion

The values of the constant β are collected from the graph of Eq. 10 shown in Figs. 2, 3 and 4. The values of β are then substituted into Eq. 11. The results were shown in Tables 1, 2, and 3 for AISI 4340 depending on different values of b (b = 0.5, 0.25, and 0.75) compared with the natural frequency from their uniform rod counterpart.

According to the theory, the natural frequency is supposed to decrease if cracks were present within the specimen. All tables above show that the value of the natural frequency calculated when a crack is present does correspond with the theory. As for the trend of the differences in the natural frequency of a cracked and a uniform rod, it could be said that the changes in natural frequencies between each selected mode exhibited an increasing trend for crack position b = 0.5. However, for crack positions b = 0.25 and b = 0.75, fluctuations could be observed in changes in natural frequency in which further study is still required.

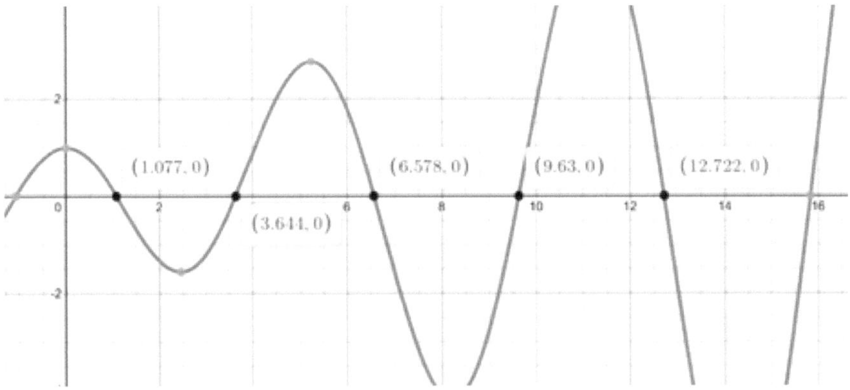

Fig. 2 Eigenfunction curve for cracked rod at crack position b = 0.5

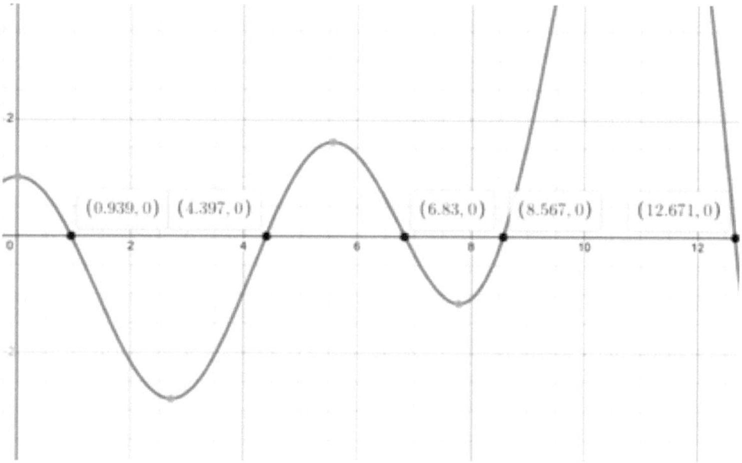

Fig. 3 Eigenfunction curve for cracked rod at crack position b = 0.25

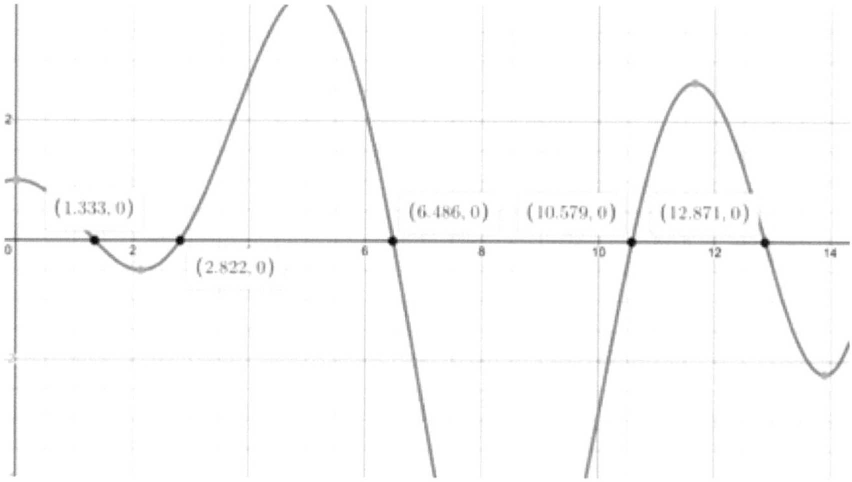

Fig. 4 Eigenfunction curve for cracked rod at crack position b = 0.75

Table 1 Calculated natural frequencies for cracked rod at crack position (b = 0.5) and uniform rod

β	ω (rad/s)	Uniform ω (rad/s)
1.077	113311.638	165264.1641
3.644	383386.8234	495792.4923
6.578	692074.2383	826320.8205
9.63	1013176.484	1156849.149
12.722	1338487.148	1487377.477

Table 2 Calculated natural frequencies for cracked rod at crack position (b = 0.25) and uniform rod

β	ω (rad/s)	Uniform ω (rad/s)
0.939	98792.59802	165264.1641
4.397	462610.2806	495792.4923
6.83	718587.2678	826320.8205
8.567	901337.7926	1156849.149
12.671	1333121.416	1487377.477

Table 3 Calculated natural frequencies for cracked rod at crack position (b = 0.75) and uniform rod

β	ω (rad/s)	Uniform ω (rad/s)
1.333	140245.5092	165264.1641
2.822	296903.8462	495792.4923
6.486	682394.8783	826320.8205
10.579	1113021.187	1156849.149
12.871	1354163.503	1487377.477

4 Conclusion

The equation obtained via the non-local elasticity theory shows that the cracked rod's eigenfunction is dependent on multiple variables. In this study, we only limited the conditions to just a normal uniform rod and a cracked rod without varying other parameters such as the non-local constant or the dimensionless stiffness parameter. The result from the calculations of the natural frequency at different crack positions also coincides with the theory that the natural frequency of a cracked rod is less than the natural frequency of a uniform rod.

References

1. Baviskar CS, Tayade RM, Patil VG (2012) Determination of failure strength of curved plate weld joint using finite element analysis. Int J Mech Eng Robot Res 1(3):22–30
2. Chaijaruwanich A, Thongthip C (2016) Failure analysis of hydraulic rotary drill rods in a limestone mine. Int J Mech Eng Robot Res 5(4):251–254. https://doi.org/10.18178/ijmerr.5.4.251-254
3. Prabhakar MS (2009) Vibration analysis of cracked beam. Thesis (MTech). National Institute of Technology Rourkela
4. Subramani T, Manivannan R, Kavitha M (2014) Crack identification in reinforced concrete beams using ANSYS software. Int J Eng Res Appl 4(6):133–41
5. Sun Y, Li X (2018) Experimental investigation of pick body bending failure. Int J Mech Eng Robot Res 7(2):184–188. https://doi.org/10.18178/ijmerr.7.2.184-188

6. Kunla N, Jearsiripongkul T, Keawsawasvong S, Thongchom C, Lawongkerd J, Saffari PR, Saffari PR, Refahati N (2023) Identification of crack location in metallic biomaterial cantilever beam subjected to moving load base on central difference approximation. Curved and Layered Struct. 10(1):20220196. Accessed from 2023 May 30
7. Li C, He L, Guo S, Zhang Y, Wu N (2015) Identification of open crack of beam using model based method. J Vibroeng 17(6):2947–6. Accessed from 30 Sep 2015
8. Hsu JC, Lee HL, Chang WJ (2011) Longitudinal vibration of cracked nanobeams using nonlocal elasticity theory. Current Appl Phys 11(6):1384–8. Accessed from 2011 Nov 1

Mechanical System Design and Reliability Analysis

Modified ArCAI for Quadrotor Under Rotor Failure

Yu Hirayama, Kenji Uchiyama, and Kai Masuda

1 Introduction

Multirotors have been utilized for delivery and inspection. A quadrotor is especially widespread because it is easy to use and inexpensive. On the other hand, the Air Accidents Investigation Branch reported that 40 unmanned aircraft vehicle (UAV) accidents caused by rotor failure happened from 2015 to 2021 [1]. The result suggests that rotor failure is a significant factor in safe UAV utilization. In particular, the rotor failure is expected to have a heavy effect on the quadrotor because its controller depends on rotor torque and thrust.

Freddi et al. [2] proposed a fault-tolerant control (FTC) system for a quadrotor controller under rotor failure that discards yaw control for the other's attitude and altitude control. Muller and D'Andrea conducted experiments with a quadrotor that lost one or two diagonal rotors based on this strategy, they indicated that the failed quadrotor could fly [3]. Moreover, the track following and safe landing control for the failed quadrotor that applied these methods was introduced [4]. These methods' premise on the failed quadrotor is controllable. However, the quadrotor under rotor failure is sometimes uncontrollable because the rotor failure changes its dynamics and control structure. Then, it requires a controllability analysis before adapting the FTC system for the quadrotor.

The degree of controllability (DOC) [5] was proposed to represent how controllable a given system is. The measure develops from the qualitative concept of controllability into quantitative controllability. The DOC is defined by the norm between the desired state and recovery region that is the set of initial states to attain the desired state. The given system is controllable if the recovery region includes the desired state. The Available Control Authority Index (ACAI) [6] and the Available

Y. Hirayama (✉) · K. Uchiyama · K. Masuda
Department of Aerospace Engineering, Nihon University, Chiba, Japan
e-mail: csyu23031@g.nihon-u.ac.jp

G. L. Conte and O. Sename (eds.), *Proceedings of 12th International Conference on Mechatronics and Control Engineering*, Lecture Notes in Mechanical Engineering,
https://doi.org/10.1007/978-981-96-6452-8_4

reduced Control Authority Index (ArCAI) [7] were proposed to analyze the controllability of failed multirotors using the DOC. These methods calculate the recovery region by the dynamics of the multirotor and assess its DOC. However, the controllability of the failed quadrotor cannot be analyzed accurately using the ACAI and the ArCAI because these criteria do not consider the gyroscopic moment generated by the yawing of the multirotor regardless of the vehicle spinning at high speed. It makes a wrong determination for the failed quadrotor flight and blocks its mission. Then, a method should be proposed to evaluate the controllability to enhance the quadrotor's utility. We modify the ArCAI criterion that extends the DOC to a failed quadrotor by considering the gyroscopic moment. The numerical simulation is conducted to confirm that the modified ArCAI calculates the controllability of the quadrotor under rotor failure.

2 Modeling of Quadrotor Under Rotor Failure

The rotor failure of a quadrotor heavily affects its total thrust and attitude, i.e., the DOC of the failed quadrotor relies on its dynamics. This section models the dynamics of a quadrotor by considering the gyroscopic moment.

Figure 1 shows the definition of the parameters of the quadrotor. $O_B - xyz$ denotes the body-fixed coordinate system and $O - XYZ$ denotes the inertial coordinate system. The origin O_B coincides with the center of mass of the quadrotor. The inertial of the propeller around the center of each rotor is explained by I_P. Each propeller is assumed to be treated as a symmetric disk about its rotation axis. This means the moment of inertia of the propeller can be approximated by the inertial of the disk on the same axis. The inertial of the body around $x-$, $y-$, and z-axes of the body coordinate system is denoted by $I_B = \text{diag}\left(I_{Bx}, I_{By}, I_{Bz}\right)$. It is assumed that the product of inertia is zero, and the total inertial of the vehicle around the body coordinate system $I = \text{diag}\left(I_x, I_y, I_z\right)$ consists of the body's moment of inertia I_B and the propeller's moment of inertia I_P of each rotor, such that $I = I_B + 4I_P$. The quadrotor is geometrical symmetry, i.e., $I_x = I_y$.

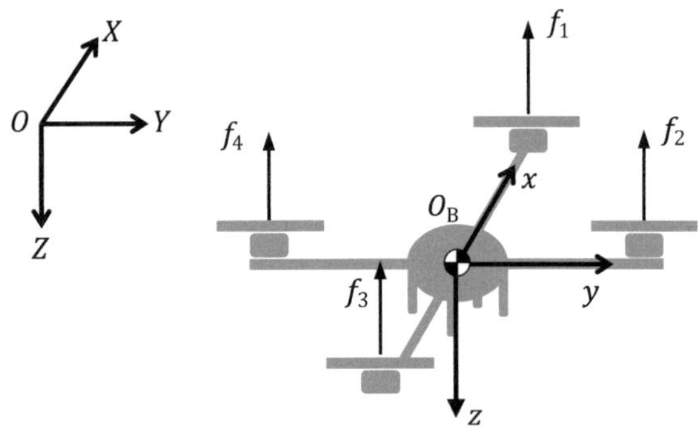

Fig. 1 Parameter's definition of quadrotor

The rotational motion of the quadrotor considering a gyroscopic moment relies on each rotor thrust and torque, and the dynamics is written by [3]

$$\begin{bmatrix} I_x\dot{p}(t) \\ I_x\dot{q}(t) \\ I_z\dot{r}(t) \end{bmatrix} = \begin{bmatrix} l(f_4 - f_2) - (I_z - I_y)q(t)r(t) - I_P q(t)\omega_\Sigma \\ l(f_1 - f_3) + (I_z - I_x)p(t)r(t) + I_P p(t)\omega_\Sigma \\ -\gamma r(t) + \kappa_\tau (f_1 - f_2 + f_3 - f_4) \end{bmatrix} \tag{1}$$

where $p(t)$, $q(t)$, and $r(t)$ represent the angle velocities of roll, pitch, and yaw, respectively; f_i and ω_i are the thrust and the torque produced by the $i-$th rotor; and the subscript Σ denotes the summation of the parameter. It is noted that the direction of the rotor rotation defines the sign of ω_i. Each thrust and torque act on the center of a rotor at a distance l from the origin O_B shown in Fig. 1. The yaw torque consists of the aerodynamic drag that is proportional to the yaw rate r with the drag coefficient γ, and the total rotor torque is the product of the constant κ_τ and $(f_1 - f_2 + f_3 - f_4)$. The yaw rate is extracted from (1), and can be solved as

$$r(t) = \left(1 - e^{-\frac{\gamma}{I_z}t}\right)\frac{\kappa_\tau}{\gamma}(f_1 - f_2 + f_3 - f_4) \tag{2}$$

The angular velocity vector satisfies $\left[p(t)\ q(t)\ r(t)\right]^T = \left[0\ 0\ 0\right]^T$. Each rotor thrust $f_i = mg/4$ in hover flight before rotor failure.

It is assumed that the quadrotor hovered before rotor failure such that the angular velocity vector at the time of rotor failure equals the condition in hover flight before rotor failure. Thus, the angular velocity vector can be represented as

$$\left[p_0\ q_0\ r_0\right]^T = \left[0\ 0\ 0\right]^T \tag{3}$$

It means that the rotor failure does not affect the angular velocity vector of the quadrotor at the time. On the other hand, rotor failure heavily affects rotor thrust, e.g., the rotor propeller stops. Each rotor thrust f_i can be expressed by using the health rate $\eta_i \in [0, 1]$ and the thrust f for each rotor as

$$f_i = \eta_i f \tag{4}$$

The rotor thrust is proportional to the angular velocity ω_i of each rotor squared as [8]

$$\omega_i = \sqrt{\frac{f_i}{\kappa_f}} \tag{5}$$

where κ_f is the ratio between the rotor thrust and the angular velocity squared. The thrust and the angular velocity of each rotor at the time of rotor failure can be solved by the condition in hover flight before rotor failure as

$$f_i = \frac{\eta_i mg}{4},\ \omega_i = \sqrt{\frac{\eta_i mg}{4\kappa_f}} \tag{6}$$

Let the recovery time and the time of rotor failure be defined t or t_0. When a rotor fails, controlling all attitude and altitude is impossible because the rotor torque generated by the healthy rotors makes the vehicle spin about the yaw axis. Then, the attitude controller should eliminate the yaw rate control from (1), i.e., the quadrotor's attitude is controlled by the roll-and-pitch moments of force. The state vector is called reduced state vector $\mathbf{F} = \left[f_\Sigma\ \tau_\phi(t)\ \tau_\theta(t)\right]^T$ with total rotor thrust f_Σ and roll-and-pitch angular moments $\tau_\phi(t) = I_x\dot{p}(t)$ or $\tau_\theta(t) = I_x\dot{q}(t)$. The roll-and-pitch angular accelerations can be solved by dividing I_x on both sides of the second and the third rows in (1). The acceleration is derived as

$$\dot{\mathbf{X}}(t) = \mathbf{A}(t)\mathbf{X}(t) + \mathbf{Bf} \tag{7}$$

where $\mathbf{X}(t) = \left[p(t)\ q(t)\right]^{\mathrm{T}}$, $\mathbf{f} = \left[f_1\ f_2\ f_3\ f_4\right]^T$ and

$$\mathbf{A}(t) - \begin{bmatrix} 0 & -\frac{\upsilon(t)}{I_x} \\ \frac{\upsilon(t)}{I_x} & 0 \end{bmatrix}, \mathbf{B} = \begin{bmatrix} 0 & -\frac{l}{I_x} & 0 & \frac{l}{I_x} \\ \frac{l}{I_x} & 0 & -\frac{l}{I_x} & 0 \end{bmatrix}$$

$$\upsilon(t) = (I_z - I_x)r(t) + I_P\omega_\Sigma \tag{8}$$

The variable υ equals the angular momentum around the z-axis. The analytical solution of (7) can be obtained as

$$\mathbf{X}(t) = \boldsymbol{\phi}(t, t_0)\mathbf{X}(t_0) + \int_{t_0}^{t} \boldsymbol{\phi}(t, \tau)\mathbf{Bf}d\tau \tag{9}$$

$$\boldsymbol{\phi}(t, t_0) = exp\left(\int_{t_0}^{t} \mathrm{A}(q)dq\right) \tag{10}$$

where $\mathbf{X}(t_0)$ represents the initial condition of the state vector and $\boldsymbol{\phi}(t, \tau)$ denotes the state transition matrix that satisfies $\boldsymbol{\phi}(t, t) = \mathbf{I}$. The solution of the state transition matrix can be solved by the analytical solution of the state transition matrix as

$$\boldsymbol{\phi}(t, t_0) = \begin{bmatrix} \cos\frac{\Upsilon(t,t_0)}{I_x} & -\sin\frac{\Upsilon(t,t_0)}{I_x} \\ \sin\frac{\Upsilon(t,t_0)}{I_x} & \cos\frac{\Upsilon(t,t_0)}{I_x} \end{bmatrix} \tag{11}$$

$$\Upsilon(t, t_0) = \int_{t_0}^{t} \upsilon(\tau)d\tau \tag{12}$$

The initial state vector is given $\mathbf{X}(t_0) = \left[0\ 0\right]^{\mathrm{T}}$. Then, (9) is rewritten as follows:

$$\mathbf{X}(t) = \int_{t_0}^{t} \boldsymbol{\phi}(t, \tau)d\tau\mathbf{Bf} \tag{13}$$

Multiplying the moment of inertia I_x by the derivative of (13) results in the following torque vector:

$$\begin{bmatrix} \tau_\phi(t) \\ \tau_\theta(t) \end{bmatrix} = I_x \frac{d}{dt} \int_{t_0}^{t} \boldsymbol{\phi}(t, \tau) d\tau \mathbf{Bf} \tag{14}$$

Thus, the torque vector can be solved as

$$\begin{bmatrix} \tau_\phi(t) \\ \tau_\theta(t) \end{bmatrix} = I_x \boldsymbol{\phi}(t, t_0) \mathbf{Bf} \tag{15}$$

It should be noted that the roll-and-pitch torques are proportional to the thrust vector, i.e., the reduced state vector $\mathbf{F} = \begin{bmatrix} f_\Sigma & \tau_\phi(t) & \tau_\theta(t) \end{bmatrix}^T$ can be defined with respect to thrust vector as

$$\mathbf{F}(t) = \mathbf{B}_f(t)\mathbf{f}, \; \mathbf{B}_f(t) = \begin{bmatrix} \mathbf{J}_{1,4} \\ I_x \phi(t, t_0)\mathbf{B} \end{bmatrix}, \; \mathbf{J}_{1,4} = \begin{bmatrix} 1 & 1 & 1 & 1 \end{bmatrix} \tag{16}$$

where $\mathbf{B}_f$ is the control effect matrix that represents how the attitude and altitude of the failed quadrotor vary with the thrust vector. It should be noted that the control effect matrix relies on rotor thrusts.

3 Modified Available Reduced Controllability Authority Index

The ArCAI calculates the recovery region of time-invariant system. However, the quadrotor dynamics under rotor failure is the time-variant system. In this section, ArCAI is extended to the failed quadrotor and calculates the DOC of quadrotor under rotor failure.

3.1 *Extending Recovery Region for Time-Variant System*

When the recovery time is divided N into equal intervals ΔT, the state vector $\mathbf{X}(t)$ in (9) can be rewritten by the following equation:

$$\mathbf{X}(k\Delta T) = \boldsymbol{\phi}(k\Delta T, k_0\Delta T)\mathbf{X}(k_0\Delta T) + \int_{k_0\Delta T}^{k\Delta T} \boldsymbol{\phi}(k\Delta T, \tau)\mathbf{B}(\tau)\mathbf{u}(\tau) d\tau \tag{17}$$

where k_0 is the initial step. The second term in (17) is defined $\Re(k\Delta T, k_0\Delta T)\mathbf{f}_k$. The reduced state vector is written as

$$\mathbf{X}(k\Delta T) = \boldsymbol{\phi}(k\Delta T, k_0\Delta T)\mathbf{X}(k_0\Delta T) + \Re(k\Delta T, k_0\Delta T)\mathbf{f}_k \tag{18}$$

where

$$\mathbf{f}_k = \begin{bmatrix} \mathbf{f}((k-1)\Delta T) \\ \vdots \\ \mathbf{f}((k_0+1)\Delta T) \\ \mathbf{f}((k_0)\Delta T) \end{bmatrix} \in \boldsymbol{R}^{N\times 4} \tag{19}$$

The matrix $\mathbf{f}_k$ is N history from k_0 to $k-1$ of the thrust vector. The reachability matrix $\Re(k\Delta T, k_0\Delta T)$ is clearly a non-negative matrix in $k > k_0$. The transition and the reachability matrices can be explained [9] as

$$\boldsymbol{\phi}(k\Delta T, k_0\Delta T) = \mathbf{A}((k-1)\Delta T)\mathbf{A}((k-2)\Delta T)\ldots\mathbf{A}((k+1)\Delta T)\mathbf{A}(k_0\Delta T) \tag{20}$$

$$\Re\big(k\Delta T, k_f\Delta T\big) = \begin{bmatrix} \mathbf{B}((k-1)\Delta T) \\ \mathbf{A}((k-1)\Delta T)\mathbf{B}((k-2)\Delta T) \\ \vdots \\ \mathbf{A}\big((k_f+1)\Delta T\big)\mathbf{B}\big(k_f T\big) \end{bmatrix} \tag{21}$$

The initial state $\mathbf{X}(k_0\Delta T)$ can be attained to $\mathbf{X}(k) = 0$ as

$$\mathbf{X}(k_0\Delta T) = -\boldsymbol{\phi}^{-1}(k\Delta T, k_0\Delta T)\Re\big(k\Delta T, k_f\Delta T\big)\mathbf{f}_k \tag{22}$$

The recovery region Ω is produced by the set of all initial states that can attain the desired state. Therefore, the recovery region can be represented by the set of (22). Comparing the reduced state vector $\mathbf{F}$ to the initial state $\mathbf{X}(k_0)$, the control effective matrix $\mathbf{B}_f(t)$ equals $-\boldsymbol{\phi}^{-1}(k\Delta T, k_0\Delta T)\Re(k\Delta T, k_0\Delta T)$. This means the recovery region of the quadrotor under rotor failure can be represented as

$$\Omega(t) = \big\{\mathbf{F}(t) \in \boldsymbol{R}^n | \mathbf{F}(t) = \mathbf{B}_f(t)\mathbf{f}, \mathbf{f} \in \big[\underline{\mathbf{f}}, \overline{\mathbf{f}}\big]\big\} \tag{23}$$

where $\underline{\mathbf{f}}$ and $\overline{\mathbf{f}}$ denote the minimum and the maximum rotor thrust vectors, respectively and n represents the dimension of the state vector. The dimension of thrust vector is explained by m. It is noted that the recovery region $\Omega(t)$ is n-dimensional set. The inequality $N \times m \geq n$ should be satisfied to avoid dimensionality deficit. Figure 2a shows the image of the recovery region in the reduced set of states space, and the projection of the recovery region onto the LM plane is shown in Fig. 2b.

Let $\mathbf{G}(t)$ be the reduced state vector in the stable condition. In the stable state vector, the angular momentums around the roll-and-pitch axes are equilibrium of (1), and total thrust can balance weight, i.e., $f_\Sigma = mg$. The stable state vector can be denoted as

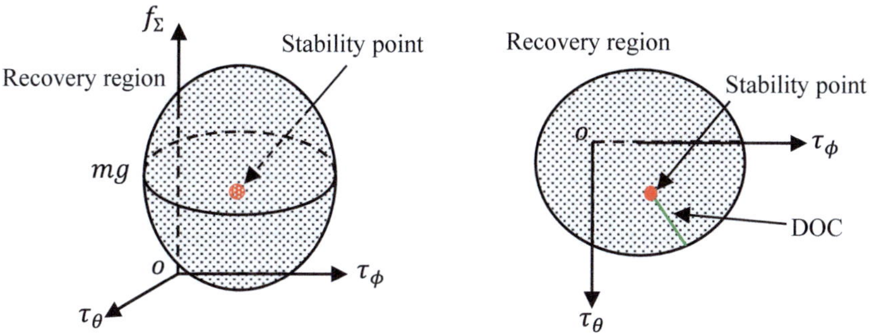

(a) Recovery region of failed quadrotor (b) Topview of recovery region and DOC

Fig. 2 Recovery region and DOC

$$\mathbf{G}(t) = \mathbf{B}_f(t)\mathbf{f}_G \tag{24}$$

with the thrust vector at stable condition $\mathbf{f}_G$. The stable state vector can be shown as the stability point in Fig. 2.

3.2 Calculating Degree of Controllability

The DOC is defined the norm between the stability point and the outer boundary of the recovery region. In the proposed method, the recovery region $\Omega(t)$ is approximated by the projections of the $(n-1)$-dimensional hyperplane segment in the control space that is expressed by thrust vector, and the calculated DOC is an approximation that is a lower bound estimate of the exact DOC. The j-th group of hyperplane segments is denoted as

$$\partial\Omega_{1,j}(t) = \left[\mathbf{B}_{1,j}(t)\ \mathbf{B}_{2,j}(t)\right]\begin{bmatrix}\bar{\mathbf{f}}_{1,j}\\ \bar{\mathbf{f}}_{2,j}\end{bmatrix} = \mathbf{B}_{1,j}(t)\bar{\mathbf{f}}_{1,j} + \mathbf{B}_{2,j}(t)\bar{\mathbf{f}}_{2,j} \tag{25}$$

$$j = 1, 2 \cdots, s,\ s = \frac{m!}{(m-(n-1))!(n-1)!} \tag{26}$$

where $\mathbf{B}_{1,j}(t) \in \boldsymbol{R}^{n\times(n-1)}$ and $\mathbf{B}_{2,j}(t) \in \boldsymbol{R}^{n\times[(N\times m)-(n-1)]}$ are corresponding partitions of the control effectiveness matrix $\mathbf{B}_f(t)$. The $n-1$ segment of rotor thrusts $\bar{\mathbf{f}}$ are selected to make $\bar{\mathbf{f}}_{1,j} \in \boldsymbol{R}^{n-1}$ and $\bar{\mathbf{f}}_{2,j} \in \boldsymbol{R}^{(N\times m)-(n-1)}$ that corresponding to $\mathbf{B}_{1,j}(t)$ and $\mathbf{B}_{2,j}(t)$. Note that s is the number of hyperplane segments group. The partition $\mathbf{B}_{1,j}(t)$ has n normal vector $\boldsymbol{\xi}_j^T(t)$. Multiplying both sides in (25) by the normal vector $\boldsymbol{\xi}^T(t)$ is

$$\boldsymbol{\xi}_j^T(t)\partial\Omega_{1,j}(t) = \boldsymbol{\xi}_j^T(t)\mathbf{B}_{2,j}(t)\bar{\mathbf{f}}_{2,j} \tag{27}$$

Equation (27) equals a hyperplane segment in the state space expressed as a reduced set of states. The distance from the center of the recovery region to the hyperplane segments and the stability point are expressed as

$$d_{B,j}(t) = \left|\boldsymbol{\xi}_j^T(t)\mathbf{B}_{2,j}(t)\bar{\mathbf{f}}_{2,j}\right|,\ d_{G,j}(t) = \left|\boldsymbol{\xi}_j^T(t)(\mathbf{F}_c(t) - \mathbf{G}(t))\right| \tag{28}$$

where $\mathbf{F}_c$ is the center of recovery region that is defined with the median thrust vector $\mathbf{f}_c$.

$$\mathbf{F}_c(t) = \mathbf{B}_f(t)\mathbf{f}_c \tag{29}$$

The distance from the stable state vector $\mathbf{G}(t)$ to the hyperplane segments in the state space can be represented as

$$d_j(t) = min\{d_{B,j}(t)\} - |d_{G,j}(t)| \tag{30}$$

The minimum distance from the center of the recovery region to the group of parallel hyperplane segments can be denoted as

$$min\{d_{B,j}(t)\} = \frac{1}{2}sign\left(\boldsymbol{\xi}_j^T(t)\mathbf{B}_{2,j}(t)\right)\boldsymbol{\Lambda}_j\left(\boldsymbol{\xi}_j^T(t)\mathbf{B}_{2,j}(t)\right)^T \tag{31}$$

where $\boldsymbol{\Lambda}_j = \mathrm{diag}\left(\bar{f}_{j,1}\cdots\bar{f}_{j,m-(n-1)}\right)$ is diagonal matrix used to the maximum thrust of each rotor. The distance from the stable state vector $\mathbf{G}(t)$ to the set of hyperplane segment is given by (28)–(31) as

$$d_j(t) = \frac{1}{2}sign\left(\boldsymbol{\xi}_j^T(t)\mathbf{B}_{2,j}(t)\right)\boldsymbol{\Lambda}_j\left(\boldsymbol{\xi}_j^T(t)\mathbf{B}_{2,j}(t)\right)^T - \left|\boldsymbol{\xi}_j^T(t)\left(\mathbf{B}_f(t)\mathbf{f}_c - \mathbf{G}(t)\right)\right| \tag{32}$$

The boundary of recovery region is configured by the group of hyperplane segments, the DOC is represented by $\rho(t)$ and using the minimum distance $d_j(t)$ as

$$\rho(t) = sign(min(d_1(t), \cdots, d_s(t)))min(|d_1(t)|, \cdots, |d_s(t)|) \tag{33}$$

Equation (33) represents the DOC of the quadrotor under rotor failure at the arbitrary recovery time. The DOC can compute the first recovery time when the initial state attains the desired state by assessing its value.

4 Numerical Simulation

This section analyzes the DOC of the quadrotor and examines the effectiveness of the previously proposed method through numerical simulation. This simulation assumes that a quadrotor has stopped the fourth rotor in this simulation. The specification of the quadrotor is shown in Table 1. The program calculates Eq. (33) using M-file [10].

Figure 3 shows the DOC of the quadrotor under rotor failure calculated by the ArCAI and the modified ArCAI. It is clear from the figure that the value of the ArCAI is zero in the numerical simulation, i.e., the criterion cannot assess the possibility of flight of the failed quadrotor. Contrarily, the DOC is positive when using the proposal method. Therefore, the failed quadrotor can be stable. The result agreed with the failed quadrotor flight [3].

Figure 4 shows that the DOC of the nominal quadrotor is 0.585. The nominal quadrotor does not cause rotor failure. It agrees with the DOC that ArCAI calculates. On the other hand, the DOC of the failed quadrotor shows oscillation. This means that the controllability and attainability change periodically as the recovery time elapses when a rotor failure occurs because of the gyroscopic moment.

Let the ratios between the second rotor thrust and the first or the third rotor thrust be ε, i.e., $f_2 = \varepsilon f_1 = \varepsilon f_3$. In the case of $\varepsilon = 0$, the second rotor does not generate thrust. Figure 5 shows that the quadrotor under rotor failure can be stable between $\varepsilon = 0$ and 2, and the DOC increases as the thrust ratio decreases. The DOC in $\varepsilon = 0$ especially agrees with the DOC of the nominal quadrotor. It suggests that if the healthy rotor generates enough thrust to hover, stopping the opposite rotor of the failed rotor is more effective than using three rotors to control the quadrotor under rotor failure. In rotor failure, the gyroscopic moment is required to equilibrate the thrust torque generated by the second rotor. It makes the recovery region smaller. Because the DOC is defined by the norm between the recovery region boundary and the stable state vector, the decreasing DOC is caused by increase of the second rotor thrust. The result agrees with the rotor thrust balance.

Table 1 Specification of the quadrotor

Parameters	Value
Mass $m[\mathrm{kg}]$	0.5
Distance between center of mass and center of rotor $l[\mathrm{m}]$	0.17
Moment of inertia $I_x, I_y[\mathrm{kg \cdot m^2}]$	3.2×10^{-3}
Moment of inertia $I_z[\mathrm{kg \cdot m^2}]$	5.5×10^{-3}
Moment of inertia of propeller $I_P[\mathrm{kg \cdot m^2}]$	1.5×10^{-5}
Ratio between rotor thrust and angular velocity squared $\kappa_f[\mathrm{N^2/rad^2}]$	6.41×10^{-6}
Ratio between rotor torque and thrust $\kappa_\tau[\mathrm{m}]$	1.69×10^{-2}
Minimum thrust of the i-th rotor $\underline{f}[\mathrm{N}]$	0
Maxim thrust of the i-th rotor $\overline{f}[\mathrm{N}]$	6

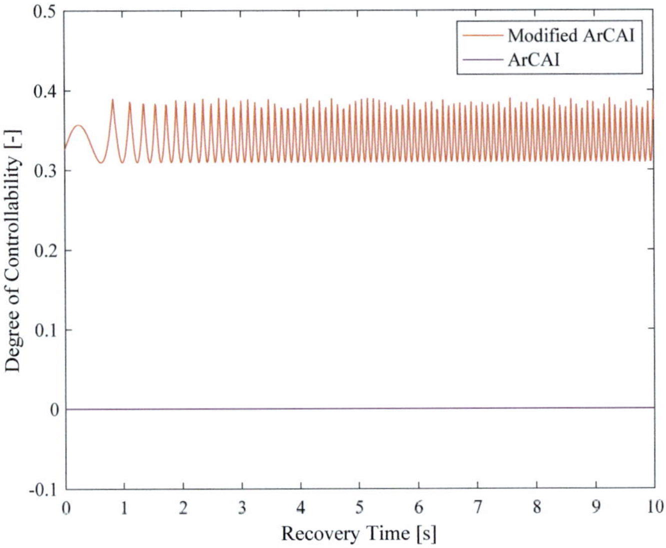

Fig. 3 The DOC of the quadrotor under rotor failure computed by modified ArCAI or ArCAI

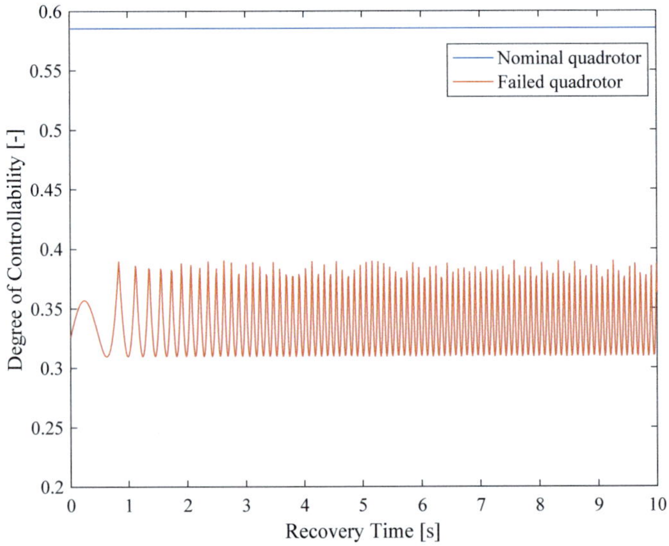

Fig. 4 The DOC of the quadrotor in nominal condition and under rotor failure

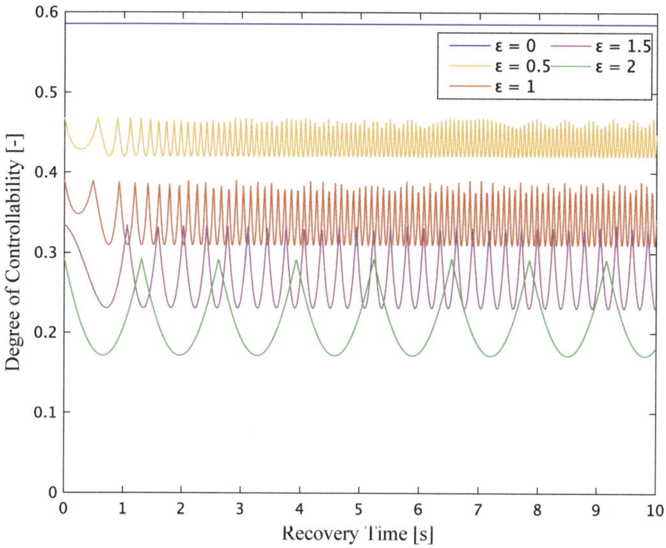

Fig. 5 The DOC of failed quadrotor which changed thrust ratio ε

5 Conclusion

We modified ArCAI by considering gyroscopic moment and proposed new method. The results of the numerical simulation conform to the quadrotor movement. It shows that the modified ArCAI successfully calculates the quadrotor's controllability under rotor failure. In the future, we will examine more safety controllers that can stabilize vehicles safely and quickly after a breakdown.

References

1. Air Accidents Investigation Branch (2022) Annual safety review 2021. AAIB, Hampshire, UK
2. Freddi A, Lamzon A, Longhi S (2011) Feedback linearization approach to fault tolerance in quadrotor vehicles. In: 18th IFAC world congress. IFAC, Milano, Italy, pp 5413–5418
3. Mueller MW, D'Andera R (2014) Stability and control of a quadrocopter despite the complete loss of one, two, or three propellers. In: 2014 IEEE ICRA. IEEE, Hong-Kong, China, pp 45–52
4. Fang N, Sihao S, Foehn P, Scaramuzza D (2022) Nonlinear MPC for quadrotor fault-tolerant control. Robot Autom Lett 7:5047–5054
5. Kalman RE, Ho YC, Narenda KN (1962) Controllability of dynamical systems. Contrib Differ Equ 1(2):189–213
6. Du G, Quan Q, Yang B, Cai K (2015) Controllability analysis for multirotor helicopter rotor degradation and failure. J Guid Control Dyn 38(5):978–985
7. Khan HIZ, Rajput J, Raiz J (2020) Reconfigurable control of a class of multicopters. In: European control conference 2020. ECC, St. Petersbrug, Russia, pp 1632–1637

8. Pounds P, Mahony R, Hynes P, Roberts J (2002) Design of a four-rotor aerial robot. In: 2002 Australasian conference on robotics and Automation, 27. ACRA, Auckland, New Zealand, p 29
9. Rumchev VG, Adeane J (2004) Reachability and controllability of time-variant discrete-time positive linear systems. Control Cybern 33(1):85–94
10. Du G, Quan Q (2016) A Matlab toolbox for calculating an available control authority index of multicopters. http://rfly.buaa.edu.cn/resources

Simulation Model Verification and Special Control Modes of DJI M600 Pro Multicopter

Peter Bauer, Mihaly Nagy, and Sandor Csurgai

1 Introduction

The goal of this article is to explore and show DJI M600 behavior in the different OSDK control modes and to verify the validity of the DJI simulator by comparing its outputs to real flight data. Both can be useful for researchers and industry practitioners who plan to develop custom onboard controllers for this drone.

Multirotor platforms with different payloads are widely applied in research and industrial applications. The DJI M600 Pro is a preferred platform due to its high payload capacity (5–6 kg) and redundant systems [2, 17, 19, 24–26, 28, 29]. Most of the applications only utilize the factory capabilities but for specific tasks custom high-level control is required as in [20, 23, 29].

A special research and development project utilizing DJI M600 Pro is Forerunner UAV development [3, 4, 13]. This means a camera-equipped multirotor UAV flying ahead of emergency ground vehicles (EGVs) to notify the driver about the hidden dangers (see Figs. 1 and 2). This task requires a special controller (see [4, 13]) as the drone should not follow rather fly ahead of the EGV knowing it's targeted next waypoint and velocity. This requires the design and application of custom controllers. This is possible through DJI Onboard Software Development Kit (OSDK) connecting a payload computer to the factory A3 autopilot through UART connection (see [3, 8]).

P. Bauer (✉)
Institute for Computer Science and Control (SZTAKI), Hungarian Research Network (HUN-REN), Kende u. 13-17.,Budapest H-1111, Hungary
e-mail: bauer@sztaki.hu; bauer.peter@sztaki.hun-ren.hu
URL: https://sztaki.hun-ren.hu/en/peter-bauer

M. Nagy
Mechatronics Engineer, Budapest, Hungary
e-mail: mihalymnagy@gmail.com

S. Csurgai
Budapest University of Technology and Economics, Budapest, Hungary

G. L. Conte and O. Sename (eds.), *Proceedings of 12th International Conference on Mechatronics and Control Engineering*, Lecture Notes in Mechanical Engineering,
https://doi.org/10.1007/978-981-96-6452-8_5

Fig. 1 Forerunner UAV ahead of an ambulance vehicle

Fig. 2 DJI M600 Pro Forerunner UAV platform with extra payload systems

OSDK makes it possible to get motion variables (position, velocity, Euler angles, angular rate) from the factory autopilot, switch different control modes, and send specific control references to the drone (see [11]). DJI also provides a simulator [9] which—connected to the drone—provides the opportunity to test OSDK-based controllers before the real flight.

Custom controllers utilizing OSDK can only be developed and tested in DJI simulator if the simulator is a valid representation of real system dynamics. So detailed comparison of DJI M600 flight responses to OSDK inputs with the DJI simulator responses to the same inputs is the goal of this paper as stated before. To the best of the author's knowledge such comparison is unpublished until now only system identification of the limited roll and pitch modes is presented in [20]. Otherwise system identification of other types of multicopters is presented, e.g., [1, 5, 6, 12, 14–16, 18, 21–23, 27]. However, detailed system identification is not always required to design controllers a validated simulation with controller test possibilities can also be satisfactory. This way this article fills the gap between the use of non-verified simulation and the detailed time-consuming identification of system dynamics.

The paper continues with the list of available OSDK modes in Sect. 2, then it shows the results of simulator verification in Sect. 3. Section 4 summarizes the special OSDK modes that behave unexpectedly from a control engineering point of view, and finally Sect. 5 concludes the paper.

2 DJI OSDK Control Modes

This section summarizes the possible OSDK control options from [11] in Table 1 and lists the targeted modes for evaluation in Table 2. There is no possibility to give mathematical and engineering details about these controllers as they are not published by DJI.

The modes listed in Table 1 can be activated by a flight control flag and X, Y, Z, and Yaw reference signals. The evaluated modes all apply HORIZONTAL_BODY considering the X-Y directions along body forward and side axes. The simulation and flight-tested maneuvers are listed in Table 2. In all cases the italic flags are the flags of the tested mode excited with the described references the other modes are commanded as listed:

1. VERTICAL_POSITION means altitude hold at the switching altitude
2. YAW_ANGLE means yaw angle hold at the switching orientation
3. HORIZONTAL_VELOCITY means zero commanded horizontal ground relative velocities and so approximate position hold at the switching position.

The stock DJI M600 was equipped with a custom onboard system to transform it into a Forerunner UAV (see Figs. 3 and 4). Figure 3 shows the connection of the onboard Nvidia Jetson Xavier NX computer to the autopilot of DJI M600 through the OSDK UART port. This makes it possible to control the drone with custom controllers sending OSDK commands through the UART port. Flight data logging for analysis and system identification is also done through OSDK, reading and saving the specific packets with 50 Hz frequency. Flight test maneuvers were executed through OSDK, and they are summarized below.

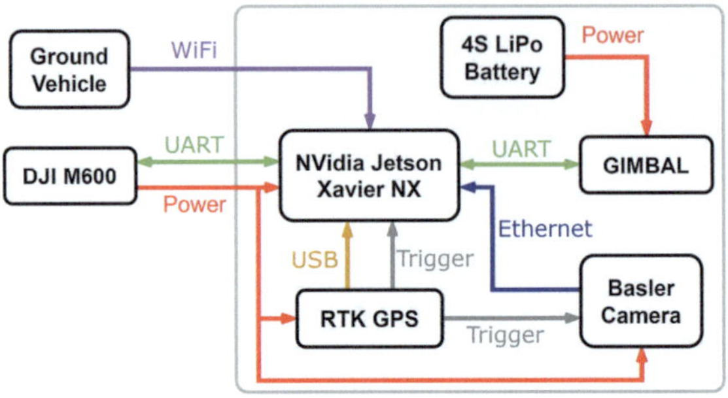

Fig. 3 Block scheme of the drone onboard hardware in the Forerunner UAV setup

The tested modes are as follows:

1. X / Y kick means a one sample position command with the given amplitude to the given direction of the body system (note that HORIZONTAL_BODY is applied).

Table 1 Possible OSDK control-mode settings

Logic identifier	Mode identifier	Explanation
HorizontalCoordinate	HORIZONTAL_GROUND	Set the x-y of ground frame as the horizontal frame (NEU)
	HORIZONTAL_BODY	Set the x-y of body frame as the horizontal frame (FRU)
	HORIZONTAL_ANGLE	Set the control-mode to control pitch & roll angle of the vehicle
HorizontalLogic	HORIZONTAL_VELOCITY	Set the control-mode to control horizontal vehicle velocities
	HORIZONTAL_POSITION	Set the control-mode to control position offsets of pitch and roll directions
	HORIZONTAL_ANGULAR_-RATE	Set the control-mode to control rate of change of the vehicle's attitude
	VERTICAL_VELOCITY	Set the control-mode to control the vertical speed of UAV, upward is positive
VerticalLogic	VERTICAL_POSITION	Set the control-mode to control the height of UAV
	VERTICAL_THRUST	Set the control-mode to directly control the thrust
YawLogic	YAW_ANGLE	Set the control-mode to control yaw angle
	YAW_RATE	Set the control-mode to control yaw angular velocity

Table 2 Executed flight test maneuvers in simulator and real flight

Maneuver	Flags
X kick ±1m, ±3m and ±5m	*HORIZONTAL_POSITION*, VERTICAL_POSITION, YAW_ANGLE
Y kick ±1m, ±3m and ±5m	*HORIZONTAL_POSITION*, VERTICAL_POSITION, YAW_ANGLE
v_T ±3m/s, ±5m/s and ±10m/s	*HORIZONTAL_VELOCITY*, VERTICAL_POSITION, YAW_ANGLE
v_N ±3m/s, ±5m/s and ±10m/s	*HORIZONTAL_VELOCITY*, VERTICAL_POSITION, YAW_ANGLE
Roll ±10°	*HORIZONTAL_ANGLE*, VERTICAL_POSITION, YAW_ANGLE
Pitch ±10°	*HORIZONTAL_ANGLE*, VERTICAL_POSITION, YAW_ANGLE
Yaw ±45° and ±90°	*YAW_ANGLE*, HORIZONTAL_VELOCITY, VERTICAL_POSITION
Yawrate ±45°/s and ±90°/s	*YAW_RATE*, HORIZONTAL_VELOCITY, VERTICAL_POSITION
Altitude ±10m	*VERTICAL_POSITION*, HORIZONTAL_VELOCITY, YAW_ANGLE
v_D ±3m/s and ±5m/s	*VERTICAL_VELOCITY*, HORIZONTAL_VELOCITY, YAW_ANGLE

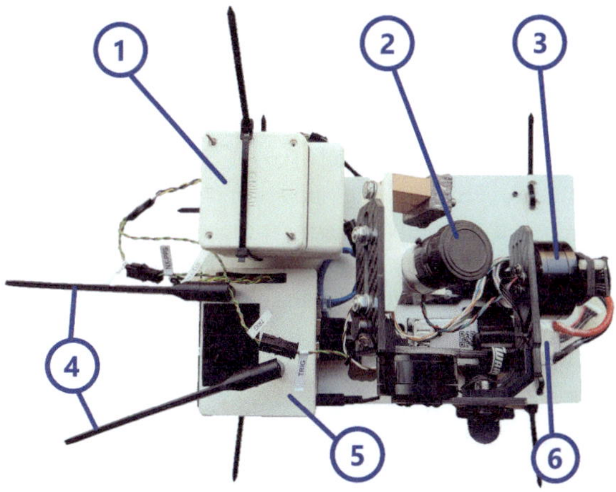

Fig. 4 Onboard hardware system of the aerial vehicle. 1: RTK GNSS electronics, 2: Basler camera, 3: Gimbal, 4: WiFi antennas, 5: Nvidia Jetson Xavier NX, 6: 4S LiPo battery

In preliminary tests it has turned out that a step position command moves the drone continuously away instead of changing its position with the given distance. This behavior is discussed in [10] and that's why only a kick command is applied for which the reaction is some movement of the drone into the given direction (for details see Sect. 4).

2. v_T means horizontal velocity reference along the X (forward) body axis with half doublet sections having zero reference in between to stop the drone.
3. v_N means horizontal velocity reference along the Y (side) body axis with half doublet sections having zero reference in between to stop the drone.
4. Roll and Pitch mean roll or pitch angle reference along the X or Y body axis. Half doublet references are applied with zero reference in between to have time for the system to reset to hover mode.
5. Yaw and Yawrate mean yaw angle or yawrate reference around the Down axis. Half doublet references are applied with zero reference in between to have time to stop drone rotation.
6. Altitude and v_D means height and vertical speed reference along the Down axis. Half doublet references are applied with zero reference in between to have time for the system to reset to hover mode. In case of altitude zero reference means the return to the initial altitude where the control code was activated.

Comparison of the simulator and flight behavior is done in the next section.

3 Verification of DJI M600 Simulator

Besides the execution of OSDK maneuvers the onboard computer of the Forerunner UAV also logs the flight data with 50 Hz frequency. GPS position, velocity, relative altitude, and Euler angles are logged. Angular rates are only logged in the factory DJI onboard log file because they are not required in the targeted control applications (position and velocity-based control see [4]).

From the listed maneuvers the X kick position control tests were not executed by OSDK, because it missed to receive the one-time step command. From the Y kick commands, only the ± 5 amplitude ones were received but this is enough to analyze system behavior and compare simulation and flight data.

In the yaw and yawrate simulation test the onboard DJI log was somehow compromised that's why there is no possibility to compare the yawrate responses of simulation and flight only the yaw angle responses can be compared.

Figures 5 to 8 show the effects of positive or negative position kick maneuvers in the side (Y) direction. The figures show that for a $5m$ (side) position kick there is about $2 - 2.5m$ displacement of the drone with a speed up and slow down maneuver giving about $2m/s$ maximum normal speeds and 20–25° maximum roll angles. Figures 5 and 6 show that the converged simulated position is a bit larger than the flight one but there is no significant difference. The normal velocity responses are also very similar (see Figs. 7 and 8). The roll angle transients are similar while the steady values are zero in the simulator (no wind disturbance) but nonzero in real flight having about $-2/-3°$ steady values in both cases (wind disturbance from the left side).

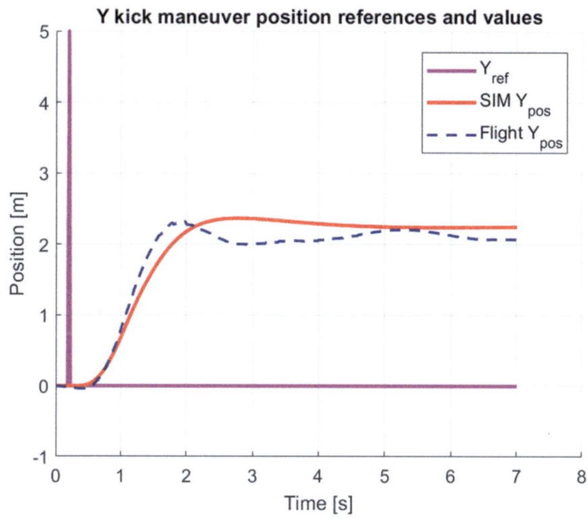

Fig. 5 Positive position kick command SIMulation and Flight position responses

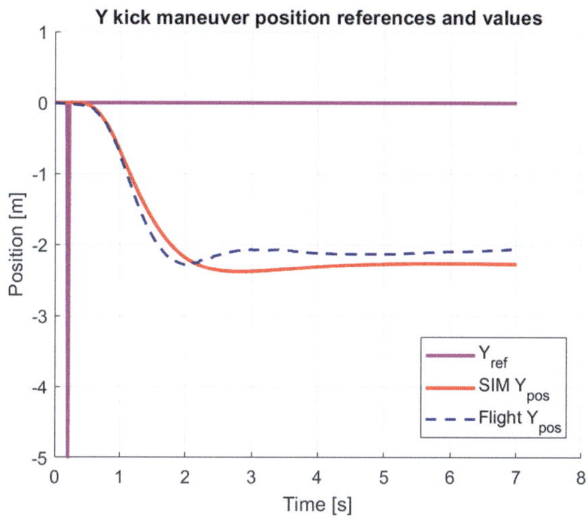

Fig. 6 Negative position kick command SIMulation and Flight position responses

Figures 9, 10, 11 and 12 show the tracking of tangential and normal velocity half doublet references with different amplitudes and the zoomed parts with the maximum $\pm 10\,\mathrm{m/s}$ amplitude. The simulation and flight test results are very close to each other regarding the velocity responses while the related pitch and roll angles have similar peak values but different steady-state ones. This is again caused by the wind disturbances.

Figures 13 and 14 show the tracking of roll and pitch half doublet references showing a special behavior detailed in the next section (Sect. 4).

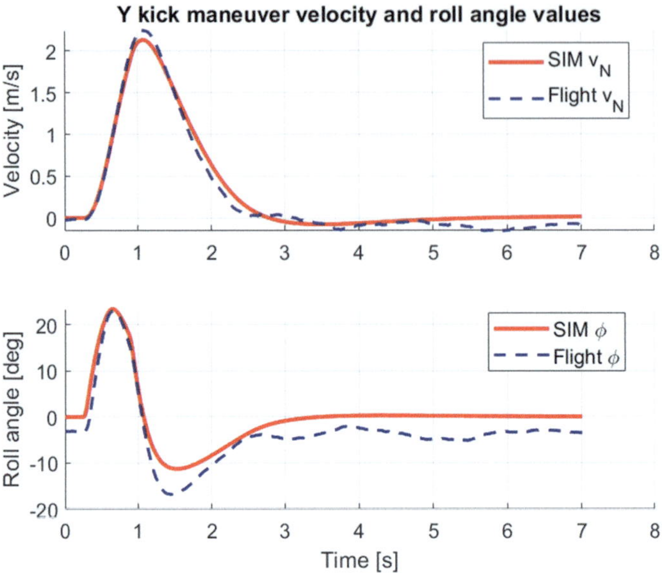

Fig. 7 Positive position kick command velocity and roll angle responses

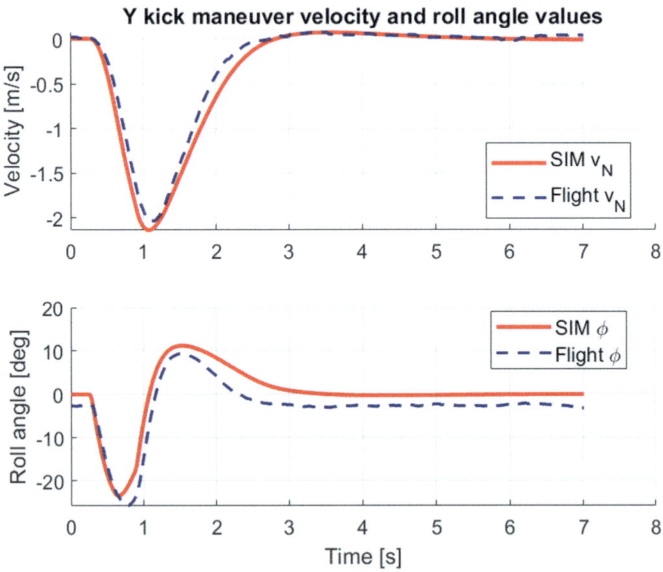

Fig. 8 Negative position kick command velocity and roll angle responses

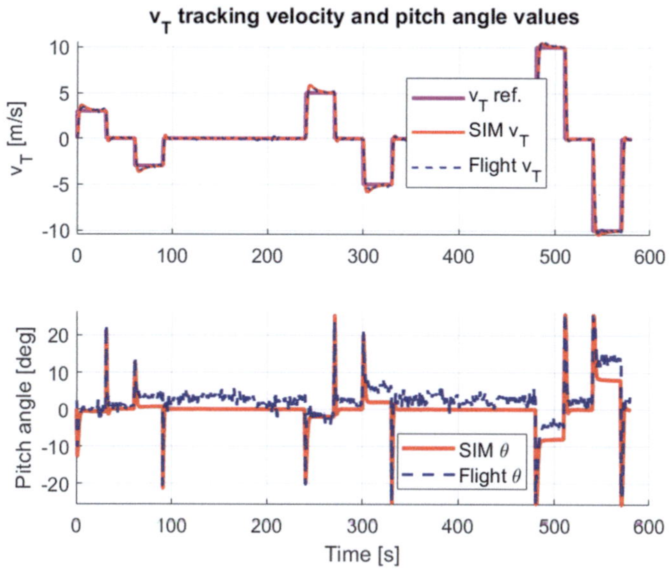

Fig. 9 Tangential velocity doublet tracking and the related pitch angles

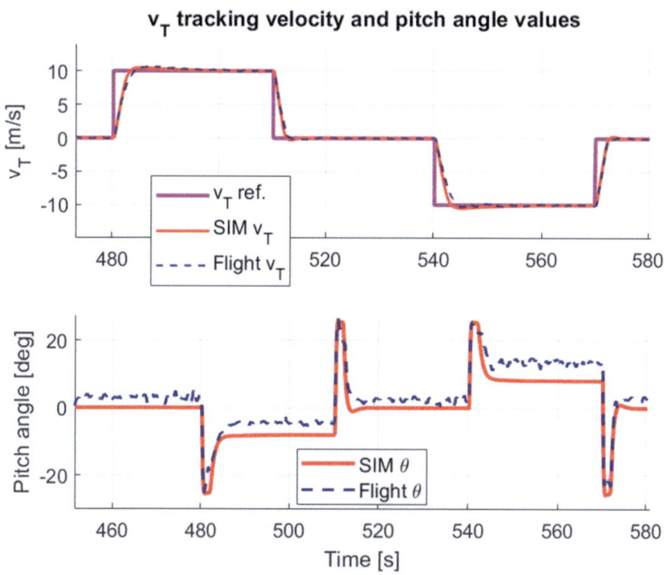

Fig. 10 Zoomed tangential velocity doublet command responses for $\pm 10\,\text{m/s}$ command and the related pitch angles

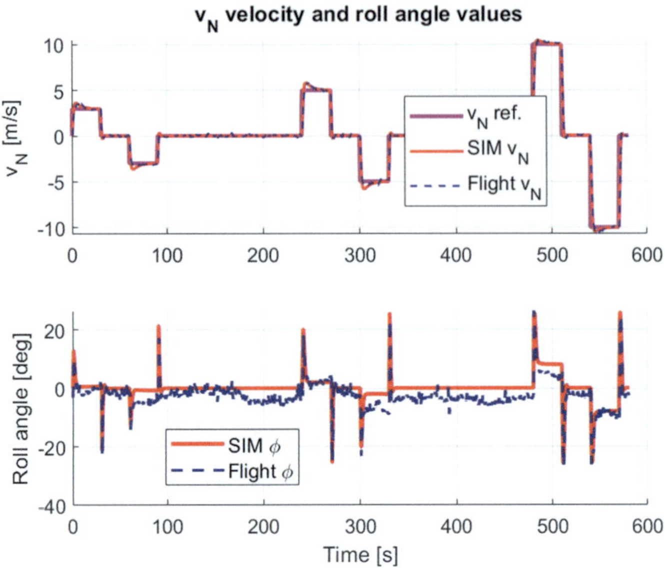

Fig. 11 Normal velocity doublet command responses and the related roll angles

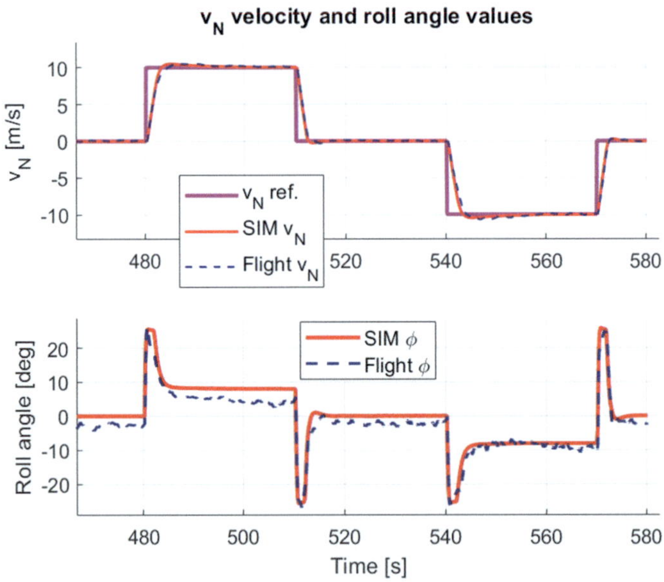

Fig. 12 Zoomed normal velocity doublet command responses for $\pm 10\,\mathrm{m/s}$ command and the related roll angles

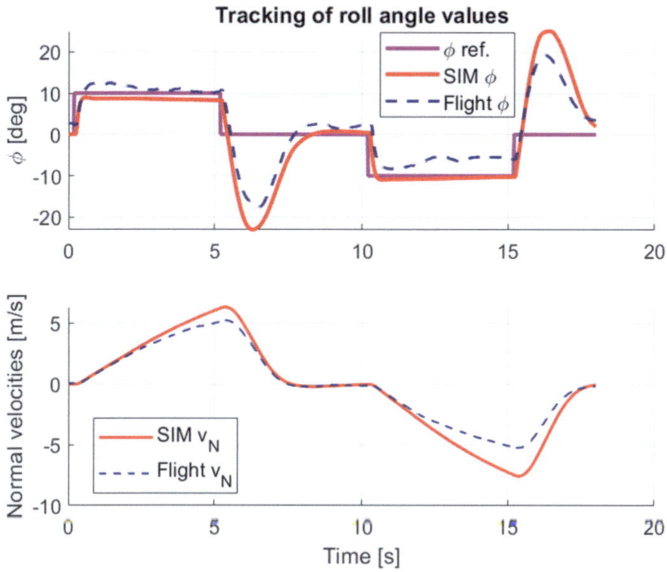

Fig. 13 Roll doublet command responses and the related normal velocities

Regarding the roll angle in Fig. 13 neither flight nor simulation values converge well to the reference and there is an offset in the steady values again caused by wind disturbance. The normal velocity well sets to zero between the two half doublets but otherwise the flight values are far from the simulated ones having a larger difference in one direction which is the direction against the wind.

In case of pitch angle tracking in Fig. 14 the behavior is similar.

Figure 15 shows the yaw angle half doublet tracking results with usually larger overshoots in flight but similar dynamics compared to the simulation.

Figure 16 shows the altitude half doublet tracking results giving almost indistinguishable curves from simulation and test flight. This is underlined by the vertical velocity tracking results (see Fig. 17) where only the transient overshoots are a bit different between simulation and flight. Note that initial altitude values are shifted to zero to make comparison possible as the simulation and flight controls were originally started at different (positive and safe) altitude values.

Figure 17 shows the tracking of vertical velocity half doublets with different amplitudes. The oscillations of the flight results are longer but the overshoots are similar. Even ± 5 m/s references are well tracked in flight and simulation both in ascend and descend contrary to the technical specifications (see [7]) stating that the maximum descent speed is limited to 3 m/s.

As a conclusion it can be stated that the DJI simulator reproduces the behavior of the drone well in every flight maneuver. The main differences are caused by the nonzero wind disturbance in real flight. Considering also the fact that nonzero wind disturbances can be applied in the simulator it is well applicable for indoor controller

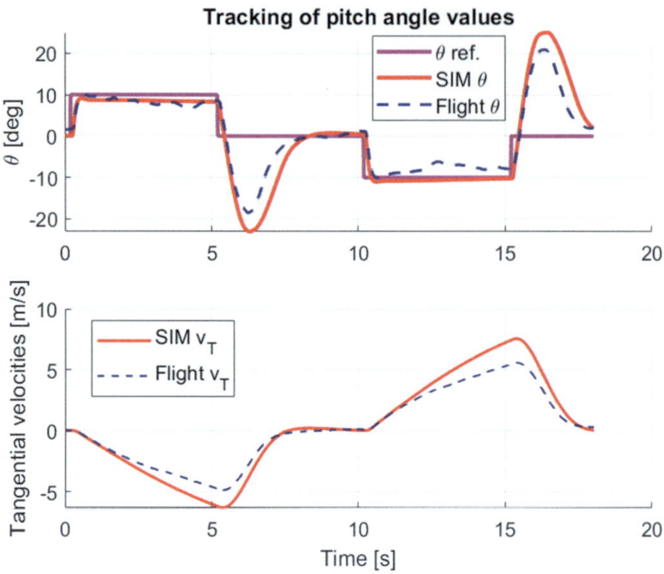

Fig. 14 Pitch doublet command responses and the related tangential velocities

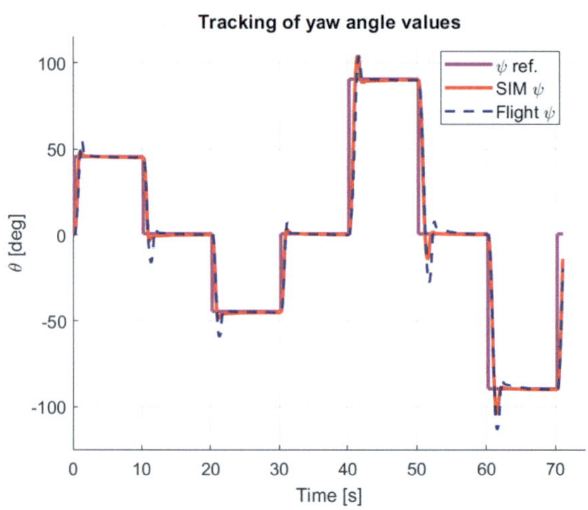

Fig. 15 Yaw angle doublet command responses

tuning and testing before real flight application. Its drawbacks are the need to connect the vehicle to the simulation computer and the need to run the control code on an embedded computer connected to its A3 autopilot. That's why drone-independent development and testing cannot be done.

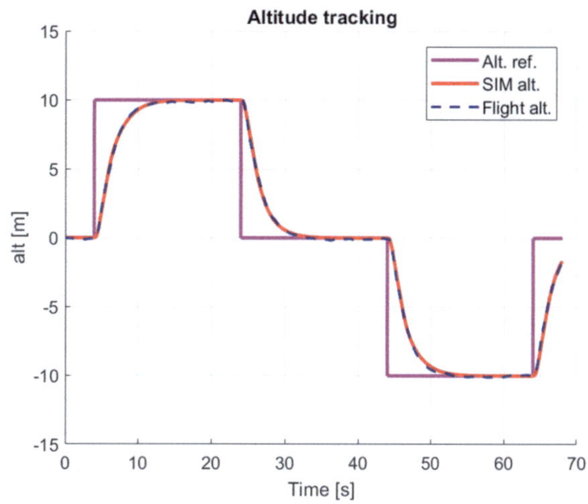

Fig. 16 Altitude doublet command responses

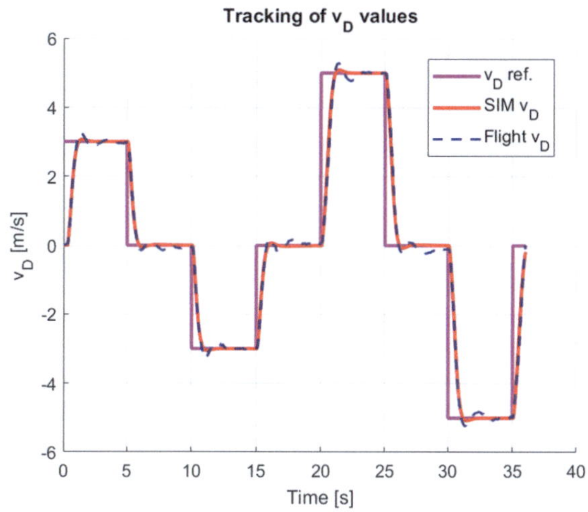

Fig. 17 Down velocity doublet command responses

4 Detected Special DJI OSDK Control Modes

During the simulation and real flight tests two unconventional modes of OSDK control were identified. One is the handling of position references, the second is the return from roll or pitch angle tracking control. For a step position command the conventional response should be the tracking of the command by moving the drone away with the given position value and then stopping. For a half doublet roll or pitch command the conventional answer should be the tracking of the positive angle reference then return to zero value. Here, conventional means that usually this is the goal with tracking control design and so the expected behavior. However, the next two subsections show that neither modes are presented in the conventional way that's why they are called unconventional.

4.1 Handling of Position Reference

It has turned out that applying a step reference to the drone position moves the drone away with a given steady velocity in the commanded direction until stopping the control action instead of moving it away with the given distance and stopping the drone at the referenced position. This behavior is described by DJI at [10]. That's why only position kick commands were applied during testing giving a position reference only for one sample (see Figs. 5 and 6). Despite only the ± 5 m amplitude Y maneuvers were successfully tested both in simulation and real flight simulation tests of $\pm$3m maneuvers were also successful. This provides the possibility to determine the maximum velocity values for a given position reference. For 5 m the maximum velocity resulted as about 2.1m/s while for 3 m as about 1.4 m/s this means about $40-50\%$ of the position reference. Considering the specialty of this mode it is better to develop a custom position tracking controller generating tangential and normal velocity references which are well followed by the drone as Figs. 9 and 11 show. This solution is applied throughout the Forerunner UAV project see [4, 13].

4.2 Return from Roll and Pitch Angle Reference Tracking

Another specialty is the handling of angle references as a step up and back to zero angle reference causes to track the nonzero value but upon jumping to the zero value a braking mode is switched to quickly stop the vehicle (see Figs. 13 and 14). The figures show that in the braking mode, the velocity quickly decreases to zero while the angles diverge far from the zero reference giving large overshoots. So zero angle reference is rather implemented as a zero velocity one stopping the vehicle.

5 Conclusion

This paper focused on two practical aspects related to the application of the DJI M600 Pro hexacopter. The first is the use of the provided DJI simulator for controller development utilizing the OSDK functionality, the second is the exploration of the special modes of OSDK control giving unexpected behavior from a control engineering point of view. Both tasks were completed by evaluating different OSDK control modes in simulation and real flight. As the practical applicability of the

HORIZONTAL_BODY mode (body referenced velocity and angle values) is better this was evaluated in all of the cases instead of the HORIZONTAL_GROUND mode (North/East referenced velocity and angle values). Except for the low-level control loops (HORIZONTAL_ANGULAR_RATE and VERTICAL_THRUST) all other modes were evaluated usually applying half doublet reference changes relative to the hover mode. The goal of this was to observe the backward transients to hover from an excited motion.

The comparison of simulator and flight data shows that the simulator well describes the dynamic behavior of the drone only the wind disturbances in real flight caused a difference. However, as there is a possibility to set steady wind disturbances in the simulator it is completely applicable for OSDK-based control development and testing. Its only drawback is the need to connect the drone to the simulation PC and so the need to implement the controller onboard the drone before testing.

Another outcome of this comparison is the observation that even 5 m/s descent speed can be followed by the system contrary to the technical specification stating only 3 m/s.

The detected special modes are the reaction for horizontal position references and the return from a horizontal angle tracking command. For a step position reference, the drone reaches a steady velocity and flies away instead of changing its position with the given value. This is because the position command is handled as a position error ‘kicking away’ the drone in the referenced direction. So finally position kick commands were applied causing only some motion of the drone. The second special mode is the return from horizontal angle (roll or pitch) tracking when the reference changes to zero. In this case the system applies a braking mode considering the zero angle reference as a zero velocity one and stopping the drone as fast as possible.

In conclusion, it can be stated that the DJI simulator is applicable for control evaluation and development but the drone should be at hand to run it and the control designer should consider the special OSDK modes (HORIZONTAL_POSITION and HORIZONTAL_ANGLE) upon designing any custom controller.

Future work included detailed identification of DJI M600 Pro dynamics resulting in a standalone Matlab/Simulink simulation model which makes control code development and evaluation faster because the drone and the onboard implementation of the control code are not required to evaluate it. The publication of this Matlab model is currently in progress together with a journal paper about detailed system identification [3].

Acknowledgements The Forerunner UAV development was financed by “Developing innovative automotive testing and analysis competencies in the West Hungary region based on the infrastructure of the Zalaegerszeg Automotive Test Track” GINOP-2.3.4-15-2020-00009.

The detailed evaluation of DJI M600 Pro capabilities was supported by the European Union within the framework of the National Laboratory for Autonomous Systems (RRF-2.3.1-21-2022-00002).

References

1. Bangura M (2017) Aerodynamics and control of quadrotors. PhD thesis, Australian National University, Canberra
2. Bates JS, Montzka C, Schmidt M, Jonard F (2021) Estimating canopy density parameters time-series for winter wheat using UAS mounted LiDAR. Remote Sens 13(4). https://doi.org/10.3390/rs13040710
3. Bauer P, Nagy M (2023) Flight-data-based high-fidelity system identification of DJIM600 Pro Hexacopter. Aerospace 11:79. https://doi.org/10.3390/aerospace11010079
4. Bauer P, Nagy M, Kuna GI, Kisari A, Simonyi E, Hiba A, Drotar I (2023) Stability focused evaluation and tuning of special ground vehicle tracking algorithms. In: Proceedings of the 22nd world congress of the international federation of automatic control (IFAC WC 2023)
5. Bouabdallah S (2007) Design and control of quadrotors with application to autonomous flying. PhD thesis, ÉCOLE POLYTECHNIQUE FÉDÉRALE DE LAUSANNE, Lausanne. https://doi.org/10.5075/epfl-thesis-3727. http://infoscience.epfl.ch/record/95939
6. Capello E, Park H, Tavora B, Guglieri G, Romano M (2015) Modeling and experimental parameter identification of a multicopter via a compound pendulum test rig. In: 2015 workshop on research, education and development of unmanned aerial systems (RED-UAS), pp 308–317. https://doi.org/10.1109/RED-UAS.2015.7441021
7. DJI (2017) DJI M600 Pro hexacopter. https://www.dji.com/hu/matrice600-pro
8. DJI (2022) DJI Onboard SDK Documentation https://developer.dji.com/onboard--api--reference/index.html. https://developer.dji.com/onboard-api-reference/index.html
9. DJI (2023) DJI assistant 2 for matrice. https://www.dji.com/hu/downloads/softwares/assistant-dji-2-for-matrice
10. DJI (2023) DJI OSDK Flight control sample. https://developer.dji.com/onboard-sdk/documentation/sample-doc/flight-control.html
11. DJI (2023) DJI::OSDK::control class reference. https://developer.dji.com/onboard-api-reference/classDJI_1_1OSDK_1_1Control.html
12. Fogelberg J (2013) Navigation and autonomous control of a hexacopter in indoor environments. Master's thesis, Lund University. Student Paper
13. Hiba A, Bauer P, Nagy M, Simonyi E, Kisari A, Kuna GI, Drotar I (2022) Software-in-the-loop simulation of the forerunner UAV system. IFAC-PapersOnLine 55(14):139–144. https://doi.org/10.1016/j.ifacol.2022.07.596
14. Ivler CM, Rowe ES, Martin J, Lopez MJS, Tischler MB (2019) System identification guidance for multirotor aircraft: dynamic scaling and test techniques. In: Vertical flight society's 75th annualforum & technology display
15. Kaya D, Kutay A (2013) Modeling, simulation, and system identification of a quadrotor helicopter. In: 7th Ankara international aerospace conference (2013)
16. Kroo IM, Shantz M, Kunz P, Fay G, Cheng SJ, Fábián T (2001) The mesicopter: a miniature rotorcraft concept phase II interim reporbt. Tech. rep., Stanford University. https://api.semanticscholar.org/CorpusID:110795824
17. Nguyen TM, Yuan S, Cao M, Lyu Y, Nguyen TH, Xie L (2021) NTU VIRAL: a visual-inertial-ranging-lidar dataset, from an aerial vehicle viewpoint. Int J Robot Res 41:270 – 280. https://api.semanticscholar.org/CorpusID:243839904
18. Niemiec R, Ivler C, Gandhi F, Sanders F (2022) Multirotor electric aerial vehicle model identification with flight data with corrections to physics-based models. CEAS Aeronaut J 13(3):575–596. https://doi.org/10.1007/s13272-022-00583-5
19. ONERA (2023) ONERA: DJI M600 application. https://www.onera.fr/fr/actualites/imagerie-laser-3d-haute-resolution-embarquee-sur-drones
20. Recalde LF, Guevara BS, Carvajal CP, Andaluz VH, Varela-Aldas J, Gandolfo DC (2022) System identification and nonlinear model predictive control with collision avoidance applied in hexacopters UAVs. Sensors 22(13). https://doi.org/10.3390/s22134712

21. Reséndiz VMA, Rivas-Araiza EA (2016) System identification of a quad-rotor in X configuration from experimental data. Res Comput Sci 118:77–86. https://api.semanticscholar.org/CorpusID:32666222
22. Rosales C, Soria CM, Rossomando FG (2019) Identification and adaptive PID Control of a hexacopter UAV based on neural networks. Int J Adapt Control Signal Process 33(1):74–91. https://doi.org/10.1002/acs.2955
23. Sa I, Kamel M, Khanna R, Popovic M, Nieto JI, Siegwart RY (2017) Dynamic System Identification, and Control for a cost effective open-source VTOL MAV. Tech. rep., Autonomous Systems Lab., Department of Mechanical and Process Engineering, ETH Zurich. https://api.semanticscholar.org/CorpusID:5737633
24. Sapkota R (2021) Using UAS imagery and computer vision to support site-specific weed control in corn. Master's thesis, North Dakota State University of Agriculture and Applied Science
25. SeaBee (2023) SeaBee Research Infrastructure: DJI M600 Pro. https://seabee.no/dji-m600-pro/
26. UNM (2023) The University of New Mexico, Center for Advancement of Spatial Informatics Research & Education: DJI M600 Equipment. https://aspire.unm.edu/facilities/equipment-computing/dji-m600.html
27. Yuksek B, Saldiran E, Cetin A, Yeniceri R, Inalhan G (2020) System identification and model-based flight control system design for an agile maneuvering quadrotor platform. In: AIAA Scitech 2020 Forum. AIAA. https://doi.org/10.2514/6.2020-1835
28. Zhang F, Song J (2018) Real-time calibration of gyro-magnetometer misalignment. IEEE Robot Autom Lett 3(2):849–856. https://doi.org/10.1109/LRA.2018.2792149
29. Zhang J, Chadha RG, Velivela V, Singh S (2019) P-CAL: pre-computed alternative lanes for aggressive aerial collision avoidance. In: Proceedings of 12th international conference on field and service robotics (FSR '19)

Modern Control Technology and Applications

Path Tracking Control for Rover with Estimation of Slip Rate Using Echo State Network

Takaki Jun, Uchiyama Kenji, and Masuda Kai

1 Introduction

Planetary exploration rover has been researched and developed as a useful tool to explore unknown planets. On an unknown planet, the land where the rover traverses is rough terrain: obstacles and uneven ground surface. Therefore, the rover needs to be guided in consideration of these existences. Obstacles can be detected with external sensors such as cameras and LiDAR. However, it is not easy to detect soft soils with them. Hence, the rover might reach soft soils and be stuck on them because of the high slip rate. When the rover has stacked on soft soils, the effect of thrust loss caused by the slip rate should be considered to move on them. Therefore, estimating the slip rate at the current position is necessary to generate the appropriate thrust. Many researches aim to estimate terrain parameters [1–5]. The slip rate was obtained in conventional methods by measuring wheel angular velocity and body velocity [1], or the image analysis [4]. Although the former method can estimate the slip rate accurately, massive sensors are required. On the other hand, the image analysis can be calculated using only a camera except for significant variations in slip rate. It is necessary to propose a flexible method of estimating large variations in the slip rate without using large-scale measurement equipment.

The estimation method of the slip rate that varies depending on the location with a few sensors is proposed. The proposed method uses the ESN to compensate for the effect of thrust loss caused by the slip rate. The ESN is one of the recurrent neural networks and can estimate some values quickly using time series data of IMU. The

T. Jun · U. Kenji (✉) · M. Kai
Nihon University, Funabashi Chiba, Japan
e-mail: uchiyama.kenji@nihon-u.ac.jp

T. Jun
e-mail: csju22021@g.nihon-u.ac.jp

G. L. Conte and O. Sename (eds.), *Proceedings of 12th International Conference on Mechatronics and Control Engineering*, Lecture Notes in Mechanical Engineering,
https://doi.org/10.1007/978-981-96-6452-8_6

ESN is used to estimate the slip rate of the current rover position in real time. Finally, the effectiveness of the proposed method is confirmed through a numerical simulation.

2 Mathematical Model

2.1 Equation of Motion

Figure 1 shows the definition of the parameters of the rover that is a crawler belt type. Its dynamics is described by translational and rotational motions [6].

$$m\ddot{x}_b = -Rsign(\dot{x}_b) + (1-\alpha)(F_l + F_r) \tag{1}$$

$$J\ddot{\psi} = -d(1-\alpha)(F_l - F_r) \tag{2}$$

where m and J are the mass and the moment of inertia of exploration rover, respectively; R denotes the coefficient of a running resistance; α denotes the wheel slip ratio, and $F_{l,r}$ is the command driving force of the exploration rover. The subscript b means a parameter of the body-fixed coordinate system. The slip ratio should satisfy the following constraint:

$$0 \leq \alpha < 1 \tag{3}$$

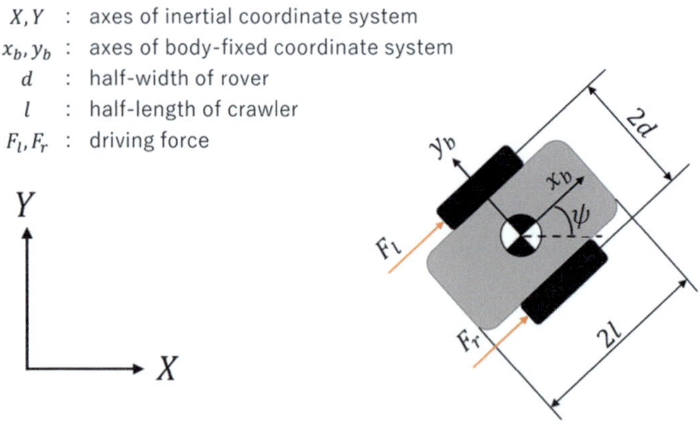

Fig. 1 Top view of rover

2.2 Linearized Model

The dynamic inversion (DI) method is employed to design a linear controller. A control input is designed to cancel the nonlinear terms to obtain a linear equation. Firstly, the position error of the rover in the inertial coordinate system is defined as

$$\mathbf{x}_e = x - \mathbf{x}_c = \begin{bmatrix} x - x_c \, y - y_c \end{bmatrix}^{\mathrm{T}} \tag{4}$$

The second time derivative of $\mathbf{x}_e$ is expressed as

$$\ddot{\mathbf{x}}_e = \ddot{\mathbf{x}} - \ddot{\mathbf{x}}_c = \ddot{\mathbf{x}} \tag{5}$$

It is noted that the vector $\ddot{\mathbf{x}}_c$ is zero because the vector $\mathbf{x}_c$ is given as a constant value. The position error $\ddot{\mathbf{x}}_e$ is translated to the body-fixed coordinate system using the translational matrix $\mathbf{C}^{B/I}$. Then, the translated position error $\ddot{\mathbf{x}}_{be}$ is written by the following equations:

$$\ddot{\mathbf{x}}_{be} = \mathbf{C}^{B/I}\ddot{\mathbf{x}}_e = \begin{bmatrix} \ddot{x}_{be} \; \ddot{y}_{be} \end{bmatrix}^{\mathrm{T}} \tag{6}$$

$$\mathbf{C}^{B/I} = \begin{bmatrix} \cos\psi & \sin\psi \\ -\sin\psi & \cos\psi \end{bmatrix} \tag{7}$$

The vector $\mathbf{x}_{be}$ is the position error in the body-fixed coordinate system. Therefore, the error equation of the rover is expressed by the following equation:

$$\ddot{x}_{be} = -\frac{R}{m}\mathrm{sign}(\dot{x}_b) - \frac{1-\alpha}{m}(F_l + F_r) \tag{8}$$

Let ψ_e be the heading angle error regarding rotational motion.

$$\psi_e = \psi - \psi_c \tag{9}$$

$$\ddot{\psi}_e = \ddot{\psi} - \ddot{\psi}_c = \ddot{\psi} \tag{10}$$

When applying the DI method for linearizing the nonlinear dynamics, the input for the total thrust U_c and the total moment of force M_c can be obtained by the following equations including the nonlinear term:

$$U_c = F_l + F_r = \frac{m}{1-\alpha}\left\{-\frac{R}{m}\mathrm{sign}(\dot{x}_b) + v_t\right\} \tag{11}$$

$$M_c = d(F_l - F_r) = -\frac{J}{1-\alpha}v_t \tag{12}$$

Here ν_t and ν_r can be considered the new control input for the linearized system. The error equation concerning the exploration rover's motion is linearized using the DI method.

$$\frac{d}{dt}\begin{bmatrix} x_{be} \\ \dot{x}_{be} \\ \psi_e \\ \dot{\psi}_e \end{bmatrix} = \begin{bmatrix} 0 & 1 & 0 & 0 \\ 0 & 0 & 0 & 0 \\ 0 & 0 & 0 & 1 \\ 0 & 0 & 0 & 0 \end{bmatrix} \begin{bmatrix} x_{be} \\ \dot{x}_{be} \\ \psi_e \\ \dot{\psi}_e \end{bmatrix} + \begin{bmatrix} 0 & 0 \\ 1 & 0 \\ 0 & 0 \\ 0 & 1 \end{bmatrix} \begin{bmatrix} \nu_t \\ \nu_r \end{bmatrix} \tag{13}$$

The linear controller is applied to decide ν_t and ν_r. This study uses the model predictive controller as the linear controller.

3 Estimation of Slip Rate

The rover gets stuck and cannot follow the reference trajectory if the slip rate estimate is different from the actual. The slip rate varies depending on the current position of the rover. Therefore, estimating the slip rate at the current position is necessary to generate the appropriate thrust. However, estimating the slip rate with external sensors such as cameras and LiDAR is difficult. In conventional methods, the slip rate was obtained by measuring wheel angular velocity and body velocity, or the image analysis. Although the former method can estimate the slip rate accurately, massive sensors are required. On the other hand, the image analysis can be calculated using only a camera except for significant variations in slip rate. It is necessary to propose a flexible method of estimating large variations in the slip rate without using large-scale measurement equipment. The proposed method uses the ESN to compensate for the effect of thrust loss caused by the slip rate. The ESN is described below [7–14].

4 Echo State Network

An ESN is a kind of artificial recurrent neural networks (RNNs). RNNs have feedback loops in their synaptic connection pathway. They can maintain an ongoing activation and thus exhibit dynamic memory. An artificial RNN can learn to estimate, predict, filter, classify, or control a complex system such as a nonlinear dynamical system. It is the most significant feature of the ESNs to update only the output weights by a simple learning algorithm. Therefore, the ESNs enable a high computing performance and quick learning with low computational load.

Figure 2 shows the ESN architecture used in this study. The subscript *ESN* means variables involved in it. The ESN consists of three units: N_u input units, N_x internal network units, and N_y output units. Activations of input units at time step n are

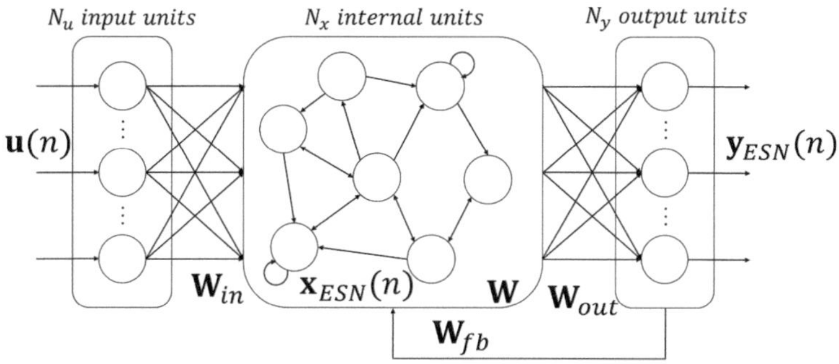

Fig. 2 The structure of the ESN considered this study

$\mathbf{u}_{ESN}(n) = \big(u_1(n), \ldots, u_{N_u}(n)\big)$, of internal units are $\mathbf{x}_{ESN}(n) = \big(x_1(n), \ldots, x_{N_x}(n)\big)$, and of output units are $\mathbf{y}_{ESN} = \big(y_1(n), \ldots, y_{N_y}(n)\big)$. Connection weights are defined as an $N_x \times N_u$ weight matrix $\mathbf{W}_{in}$ for the input weights, as an $N_x \times N_x$ matrix $\mathbf{W}$ for the internal connections, as an $N_y \times N_x$ matrix $\mathbf{W}_{out}$ for the connections to the output units, and as an $N_x \times N_y$ matrix $\mathbf{W}_{fb}$ for the connections that project back from the output to the internal units.

The activation of internal units is updated according to

$$\begin{aligned}\mathbf{x}_{ESN}(n) &= (1-\alpha_{ESN})\mathbf{x}_{ESN}(n-1)\\ &\quad + \alpha_{ESN}\mathbf{f}\big(\mathbf{W}_{in}\mathbf{u}(n) + \mathbf{W}\mathbf{x}_{ESN}(n-1) + \mathbf{W}_{fb}\mathbf{y}_{ESN}(n-1)\big)\end{aligned} \tag{14}$$

where $\mathbf{f}$ is the internal unit's output function and α_{ESN} is leaking rate. In this study, the hyperbolic tangent function is applied to the internal unit's output function $\mathbf{f}$. Leaking rate can control the speed of changes of the activation in internal units $\mathbf{x}_{ESN}$. The output is computed according to the following equation:

$$y_{ESN}(n) = \mathbf{W}_{out}\mathbf{x}_{ESN}(n) \tag{15}$$

The objective of training is to find a set of the output weights $\mathbf{W}_{out}$ such that the following squared error

$$E(n) = \tfrac{1}{2}||\mathbf{e}(n)||_2^2 = \tfrac{1}{2}||\mathbf{d}(n) - \mathbf{y}_{ESN}(n)||_2^2 = \tfrac{1}{2}||\mathbf{d}(n) - \mathbf{W}_{out}\mathbf{x}_{ESN}(n)||_2^2 \tag{16}$$

where $\mathbf{d}$ is the training data for the ESN. The function $E(n)$ in (16) is minimized by incrementally changing the output weights along the direction of the error gradient relating to weights:

$$\frac{\partial E(n)}{\partial \mathbf{W}_{out}(n)} = \frac{\partial E(n)}{\partial w_{i,j}(n)} \tag{17}$$

Using a learning rate η:

$$\mathbf{W}_{out}(n+1) = \mathbf{W}_{out}(n) - \eta \frac{\partial E}{\partial \mathbf{W}_{out}} = \mathbf{W}_{out}(n) + \eta \mathbf{e}(n)\mathbf{x}_{ESN}^{\mathrm{T}} \tag{18}$$

The celebrated least-mean square (LMS) algorithm is used for the optimization of the output weights $\mathbf{W}_{out}$ [11, 12].

4.1 Applying Constraints to the ESN

The slip rate should satisfy (3). However, the ESN's output is unconstrained. Therefore, the barrier method was applied so that the slip rate satisfies (3) [13, 14]. Barrier methods are motivated by minimizing a composite function that reflects the original objective function as well as the influence of the constraints. Here, the nonlinear programming problem is considered below:

$$\min_{\mathbf{x} \in \mathcal{R}^n} f(\mathbf{x}) subject\, to c(\mathbf{x}) \geq 0 \tag{19}$$

where $\mathbf{c}(\mathbf{x})$ has m component functions, and f and $\{c_i\}$ are smooth. The logarithmic barrier function associated with the inequality in (19) is defined as

$$B(\mathbf{x}, \mu) = f(\mathbf{x}) - \mu \sum_{j=1}^{m} \ln c_j(\mathbf{x}) \tag{20}$$

where μ is a positive scalar called the barrier parameter. The logarithmic terms are well defined at points $\mathbf{x}$ for which $\mathbf{c}(\mathbf{x}) > 0$, but become unbounded above as $\mathbf{x}$ approaches any point where a constraint is zero, and are undefined if $c_j(\mathbf{x}) < 0$ for any j. Therefore, this constrained optimization problem can be solved as an unconstrained optimization problem.

5 Numerical Simulation

5.1 Simulation Conditions

The slip rate varies depending on the current position of the rover, and roughly continuous. Therefore, the slip rate in the simulation environment was defined as the following equation:

$$\alpha(x, y) = \frac{\alpha_{max}}{\beta_x (X - x)^2 + \beta_y (Y - y)^2 + 1} \tag{21}$$

Table 1 Parameters of slip rate

Parameters		Values
α_{max}	[-]	0.9
β_x	[-]	1.5
β_y	[-]	1
X	$[m]$	0.3
Y	$[m]$	0.7

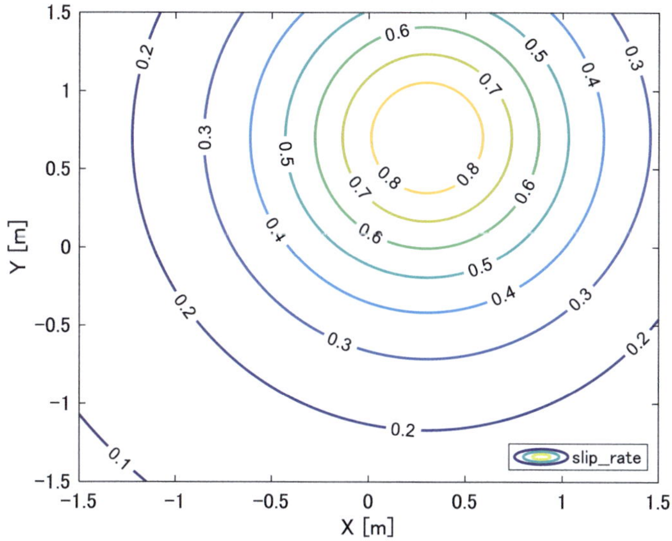

Fig. 3 Slip rate in the simulation environment

where α_{max} is the maximum of slip rate in the environment, $\beta_{x,y}$ is weights regarding change rates of slip rate, and X, Y are the location of α_{max}. Respective parameters used in the numerical simulation are indicated in Table 1. Figure 3 shows the distribution of the slip rate in XY-plane.

5.2 *Simulation Results*

In this simulation, the rover starts at position (1, 0) in the XY-plane and follows a circular trajectory counterclockwise in the environment where the slip rate varies depending on the position in the XY-plane. The ESN estimates the slip rate of the current rover position sequentially, and the rover's inputs are determined considering it. Figure 4 shows comparisons between a slip rate estimated by the ESN and a constant slip rate. Although the rover cannot follow the reference trajectory shown in Fig. 4b, the proposed method enables the rover to follow the reference trajectory

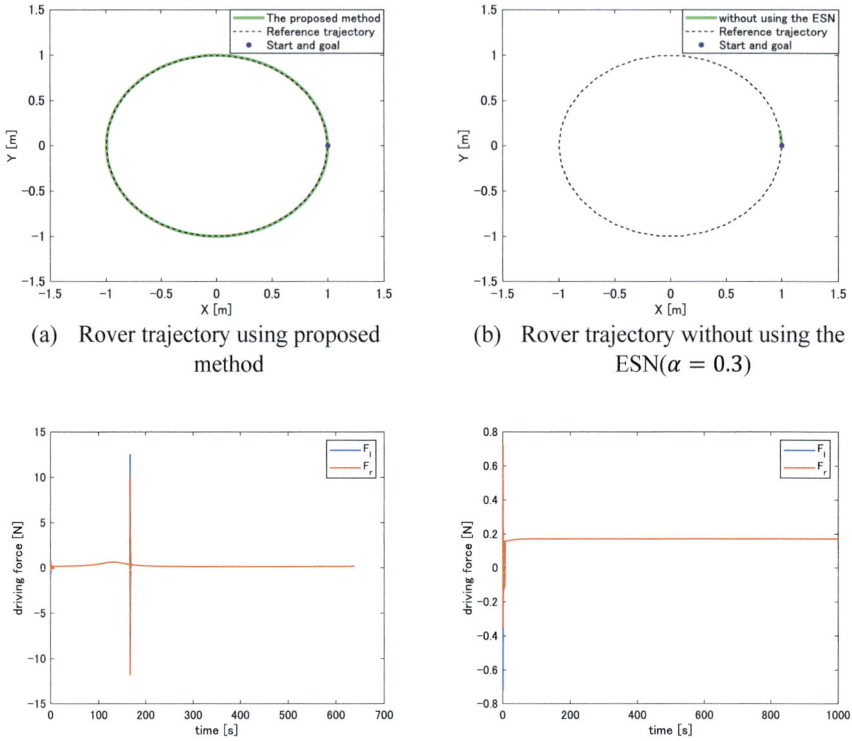

(a) Rover trajectory using proposed method

(b) Rover trajectory without using the ESN($\alpha = 0.3$)

(c) Control input using proposed method

(d) Control input without using the ESN($\alpha = 0.3$)

Fig. 4 Comparisons between the slip rate estimated by the ESN and a constant slip rate

in Fig. 4a. Driving forces of the proposed method in Fig. 4c are bigger than the forces without using the ESN in Fig. 4 d. Therefore, it is determined that the effect of thrust loss is compensated by the slip rate estimated by the ESN. Figure 5 shows the time history of slip rate estimated by the ESN. It is shown that the ESN could estimate slip rate quickly and accurately in real time.

6 Conclusions

We proposed a planetary rover tracking control method with estimating the slip rate based on the ESN. The ESN enabled the rover to follow the reference trajectory on the environment with soft soils by estimating the slip rate of the current rover position in real time. It was confirmed that the ESN can estimate it quickly and accurately by the numerical simulation. In this study, a sinkage of the rover wheels was not considered. In addition to this theoretical extension, tests using an experimental rover should be conducted in future work.

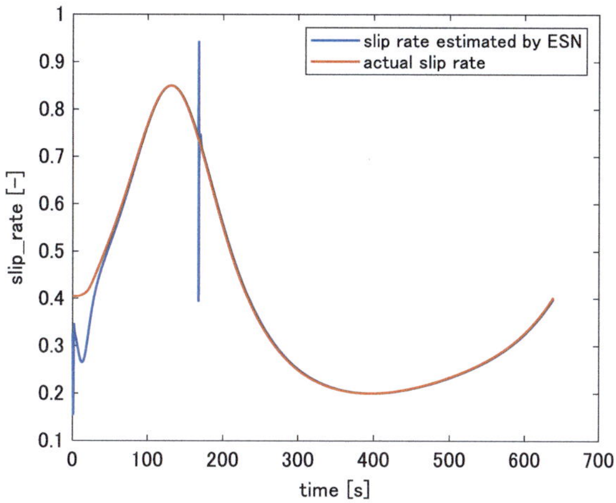

Fig. 5 Time history of the slip rate estimated by the ESN

References

1. Iagnemma K, Shibly H, Dubowsky S (2002) On-line terrain parameter estimation for planetary rovers. In: IEEE international conference on robotics and automation, pp 3142–3147
2. Ojeda L, Cruz D, Reina G, Borensten J (2006) Current-based slippage detection and odometry correction for mobile robots and planetary rover. IEEE Trans Robot 22(2):366–378
3. Nagatani K, Yamana K, Ishigami G, Yoshida K (2006) On-line estimation of slip-ratio for exploration rover on loose sand—development of wheeling with built-in force sensor array. Tohoku Univ., (in Japanese)
4. Arai Y, Iizuka K, Suzuki S, Kawamura T, Kubota T (2013) Estimation of slip ratio using information of ground for lunar exploration rovers. In: JSME conference on robotics and mechatronics 1A1-R17 (2013) (in Japanese)
5. Bussmann K, Meyer L, Steidle F, Wedler A (2018) Slip modeling and estimation for a planetary exploration rover: experimental results from Mt. Etna. In: IEEE/RSJ international conference on intelligent robots and systems, pp 2449–2456
6. Chiba S, Uchiyama K, Masuda K (2019) Guidance method without terrain information for an exploration rover. In: Australian and New Zealand control conference, pp 36–41
7. Jaeger H (2001) The “echo state” approach to analysing and training recurrent neural networks—with an Erratum note. German national research center for information technology GMD technical report 148(34):13
8. Jaeger H, Haas H (2004) Harnessing nonlinearity: predicting chaotic systems and saving energy in wireless communication. Science 304(5667):78–80
9. Jaeger H, Lukosevicius M, Popovici D, Siewert U (2007) Optimization and applications of echo state networks with leaky-integrator neurons. Neural Netw 20(3):335–352
10. Gallicchio C, Micheli A (2011) Architectural and Markovian factors of echo state networks. Neural Netw 24(5):440–456
11. Behrouz FB (2013) Adaptive filters: theory and applications, 2nd edn. Wiley, USA
12. Widrow B, Hoff ME (1960) Adaptive switching circuits. IRE WESCON Convention Record, pp 96–104
13. Conn AR, Gould N, Toint PL (1997) A globally convergent Lagrangian barrier algorithm for optimization with general inequality constraints and simple bounds. Math Comput 66(217):261–288
14. Freund RM (2004) Penalty and barrier methods for constrained optimization. Massachusetts Institute of Technology University

15. John MG, Sherif H, Shady AM (2022) Path tracking control based on an adaptive MPC to changing vehicle dynamics. Int J Mech Eng Robot Res 11(7):535–541
16. Jing Q, Zhixiang L, Youmin Z (2018) Gain scheduling based PID control approaches for path tracking and fault tolerant control of a quad-rotor UAV. Int J Mech Eng Robot Res 7(4):401–408

Advanced Motion Planning Method for Unmanned Vehicles to Consider 3D Objects

Dénes Tompos, Tamás Hegedűs, and Balázs Németh

1 Introduction and Motivation

This paper proposes an idea for connecting, motion planning algorithms for unmanned vehicles with a reinforcement learning (RL) based velocity control algorithm for optimal control and collision avoidance. To solve the problem of optimal path planning these algorithms often use map or graph of the operating space, see e.g., [1, 2]. There are existing solutions to solve the problem of a dynamic environment in the path planning, as an extension, which continuously redefines the path locally [3, 4].

Various methods have also been developed for trajectory planning and velocity control of a UAV [5]. For example, a path planning controller was designed in this paper [6], based on a triple pendulum behavior.

The proposed solution of this paper is based on previous results on route planning, can be found in [7]. In that paper a velocity control of an unmanned road vehicle is designed such a way it avoids collision with other moving vehicles. The scope of this novel paper is to provide an extension to the algorithm, which is capable of selecting

D. Tompos (✉) · T. Hegedűs · B. Németh
Institute for Computer Science and Control (SZTAKI), Hungarian Research Network (HUN-REN), Kende utca, Budapest, Hungary
e-mail: denes.tompos@sztaki.hun-ren.hu

T. Hegedűs
e-mail: tamas.hegedus@sztaki.hun-ren.hu

B. Németh
e-mail: balazs.nemeth@sztaki.hun-ren.hu

D. Tompos · B. Németh
Department of Control for Transportation and Vehicle Systems, Budapest University of Technology and Economics, Stoczek utca, Budapest, Hungary

G. L. Conte and O. Sename (eds.), *Proceedings of 12th International Conference on Mechatronics and Control Engineering*, Lecture Notes in Mechanical Engineering,
https://doi.org/10.1007/978-981-96-6452-8_7

conflict points from the pathways of the moving objects and the grid, which is the discrete representation of the moving objects around the unmanned vehicle.

The algorithm described in this paper solves the following tasks to create a conflict point for the velocity control. It selects the occupied area around the pathway of all vehicles in the system, and create lists of points from the grid. The intersection of these lists results in conflict points where collision may occur. The algorithm separates these points in different clusters based on they distance, since it is expected to have multiple conflict situations in the space. From these clusters a reduced set of the conflict points to the motion control algorithm is transferred.

2 Motion Planning and Moving of the Unmanned Vehicle

This section presents the method of motion planning for the unmanned vehicle and moving along that. It is divided into three tasks, such as building-up the environment, selecting the conflict points and designing motion profile for the unmanned vehicle.

Building-up the model of environment

To train and test the algorithm, a simulation is created. The simulation requires the modeling of the environment for automated vehicles, related to a real demonstration laboratory of Institute for Computer Science and Control (SZTAKI). Due to space limitations in the laboratory for experimenting with real vehicles, a simulation environment measuring 4m by 3m by 2m is created instead. In this space test, a pathway is planned for a mobile robot that moves in the simulated space in an elliptical shape and serves as a mobile obstacle in the simulation.

Moreover, a pathway is also created for the unmanned vehicle, i.e., drone in this example, which moves parallel to the y axis and crosses the path of the mobile robot. It is also necessary to avoid a static wall, placed parallel to the x axis with a height of 1m. In this example the collision to the wall is avoided through the vertical motion of the drone, while in case of the mobile robot the velocity of the drone is modified to avoid collision. The built-up environment model is illustrated in an example, see Fig. 1a. It shows the paths followed by the objects in the system. The path of the drone is marked with green and the path of the car is marked with blue.

The pathways are defined as a list of point which the vehicles have to pass trough. In simulation the objects move toward its next position in this list. The next position is determined by the distance the vehicles travelled.

Conflict point selection method

The goal of the conflict point selection is to find the conflict points in the space, where the pathways of the moving agents meet, and thus, the space occupied by the moving vehicles overlaps. For this purpose the paths of the moving objects are marked on the grid of a graph.

In the proposed algorithm for every point of the path P are dedicated as graph G points within a predefined safe distance s_{safe} designated to the vehicle.

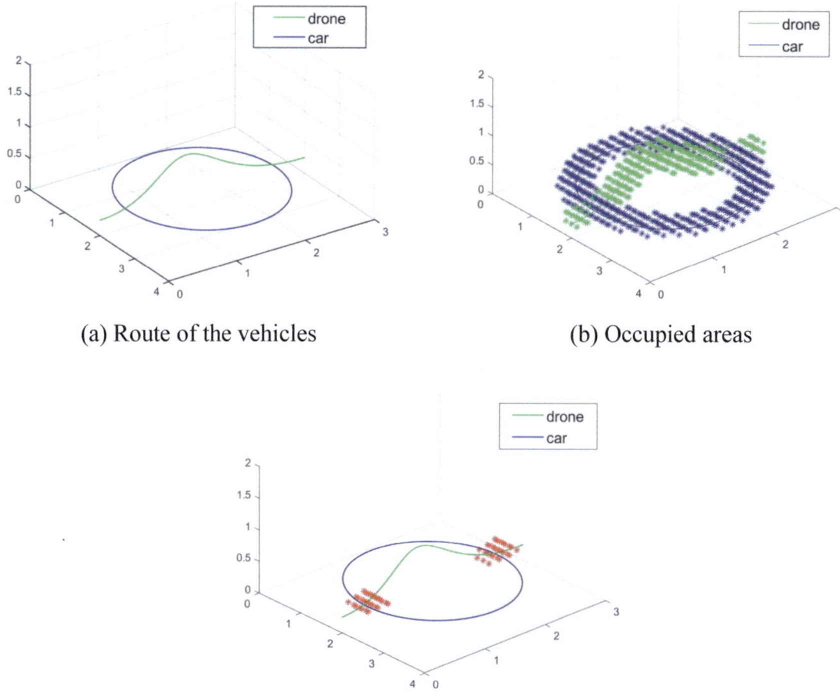

(a) Route of the vehicles

(b) Occupied areas

(c) Clusters for conflict point selection

Fig. 1 Illustrative example on the operation of the method

This safe distance represents the physical dimensions of the vehicles, which is defined as a sphere and its position is in the center P_0. Consequently, all the graph points P are selected that the vehicle can occupy, such as $\forall P \in G, P_0 \in P \| P_0(x, y, z) — P(x, y, z) \| < s_{safe}$, see Algorithm 1. The difficulty of this solution is that high computation effort is required. But from the viewpoint of the simulation, the graph is calculated outside of the simulation environment. Moreover, if the coordinates marking the path are scares, it is not necessary to select all graph points. In the given example, Fig. 1b illustrate the paths followed by the objects in the system and the space occupied by them around the pathways. The path of the drone is marked with green and the path of the car is marked with blue.

The result of the selection is a list, containing the data for all the conflict points, which should be separated with a clustering method. The method used in this case is based on the distance between the points in the list. Define a cluster as a subgroup of intersecting points where these points will not have a safe point between any of they two members where the controlled vehicle can safely stop. Thus a cluster of points can be separated with selecting points that are within safe distance from each other. During the proposed method it is necessary to select the first point from the list and check, if there are other points within s_{safe}. If yes, the neighborhood of those points is also checked, until no further points within safe distance is found. Separating one cluster from the list of conflict points. This process is repeated until all points are separated into clusters. Using the separated clusters, the conflict points are determined as the mean of these clusters, and the range of the circles are the distance of the furthest point of each cluster [8].

Algorithm 1 Clustering method for conflict point selection

$conflictpoints = P_1, P_2, \ldots, P_n$
while $conflictpoints \neq \emptyset$ **do**
  $cluster_k \leftarrow P_1$
  for $\forall P_c \in cluster_k$ **do**
    if $\forall P \in conflictpoints, P_c \in cluster \| P_c(x,y,z) - P(x,y,z) \| < s_{safe}$ **then**
      cluster $\leftarrow P$
    end if
  end for
end while=0

Figure 1c illustrates for the clusters of conflict points around the conflict situation in the given example. The path of the drone is marked with green and the path of the car is marked with blue and the conflict point are red.

Method for designing motion profile

The motion of the drone is performed with a control algorithm, which has been presented in a previous work [9]. The aim of the method is to consider interactions of unmanned vehicles when crossing each others routes with assuming both vehicles move straight. One of the unmanned vehicle must adapt to the motion of the other vehicle in such a way that a minimum safe distance from it is kept at any time of the interaction. The goal of this control is to maximize the speed of the controlled vehicles, while simultaneously to avoid collision with the surrounding objects.

The structure of the motion profile design is illustrated in Fig. 2. It contains a learning-based agent and a supervisor. The supervisor computes a $u(k)$ acceleration input for the drone in each k time instance, with which collision to the surrounding objects can be avoided. The learning-based agents provide a candidate acceleration signal to the supervisor. In this work it is designed using a reinforcement-learning-based algorithm, which attempts to optimize the energy consumption of the system via minimizing the acceleration and deceleration of the drone. Thus, the learning-based agent and the supervisors work together: the agent facilitates energy and time performance requirements, the supervisor guarantees collision avoidance.

The energy and time optimal motion through the reinforcement learning is achieved, via the appropriate definition of reward. During the training process the agent receives a reward from the environment, which reward reflects to the selected candidate acceleration. This reward is defined arbitrary, based on the expectation what the agent supposed to be learning, since it is attempted to maximise its reward value in the learning process. The agent must learn to move the controlled object

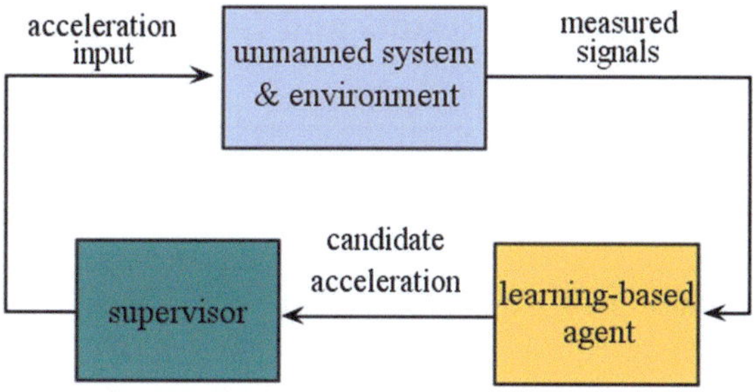

Fig. 2 Structure of the route planning and tracking system

forward and optimize it energy consumption, i.e., it is rewarded for gaining distance with minimum amount is acceleration. Moreover, reward is lost if the two conflicting object enters the avoidable circle the same time. The reward function is defined as:

$$r(k) = d(k) - Q1|u(k)| - \Theta(k) \tag{1}$$

where d is the distance between the drone and the conflict point. The magnitude of acceleration input $u(k)$ is applied in the reward with a constant penalty weight $Q_1 >$ 0. Another penalty $\Theta(k)$ is in relation with the entering of the drone into the conflict zone, i.e., its value is 1 if it is in the zone, otherwise 0. During each episode the reward is accumulated, which means that $\sum_{k=0}^{N} r(k)$ is used for the evaluation of each episode.

The formulation of the algorithm in the supervisor is based on a constrained quadratic optimization process. The goal of the supervisor is to guarantee safe drone motion and thus, e.g., Vehicle 1 (drone in the example) and Vehicle 2 (mobile robot) cannot be in the conflict area $\mathrm{R}(s_{safe})$ in the same time. It forms the safety constraint in the algorithm of the supervisor: $(S_1(k+1), S_2(k+1)) \in/\mathrm{R}(s_{safe})$, where $(S_1(k+1), S_2(k+1))$ are predicted distances at $k+1$ time step. If this constraint can be guaranteed by $u_L(k)$, then $u(k) = u_L(k)$ can be applied. But, if this constraint is violated by $u_L(k)$, then $\Delta(k)/= 0$ must be selected within the range of Δ_{bound} and thus, $u(k)$ is able to guarantee safe motion. Consequently, the supervisory algorithm is formed as:

$$\min_{\Delta(k)} \Delta(k)^2 \tag{2a}$$

subject to

$$\Delta(k) \in \Delta bound \tag{2b}$$

$$(S1(k+1), S2(k+1)) \in \mathrm{R}/(ssafe). \tag{2c}$$

3 Simulation Results and Conclusions

In the simulation example the drone has been controlled successfully to avoid collision to an obstacle road vehicle in the environment. In the example the drone moves on its route forward and backward, while the vehicle moves on its circle-shaped route, see Fig. 1a. Figure 3 shows conflict situations in the simulation. The drone velocity in Fig. 3a shows that the drone has to stop sequentially to avoid collision to the another vehicle. Trajectories of the vehicles during the simulation can be found in Fig. 3b, i.e., the collision is avoided. It is also confirmed through the illustration on 3c, where S_1, S_2 represent the distances of both vehicles to the actually selected conflict point. The illustration shows that $s_{safe} = 0.3\ m$ can be guaranteed during the motion.

The optimization problem during the motion of the UAV must be continuously solved, and thus, the possibility of real-time computation must be achieved. *quadprog* is the built-in solver of the MATLAB [10] was used for the optimization. Throughout the simulation tests, the optimization time is recorded. The total length of the simulation is 100 s and the sampling time is set to 0.1 s. The average time of computing the results is 0.002 s with the computation time fluctuating between 0.0018 s and 0.0039, thus the standard deviation is 0.00017 s.

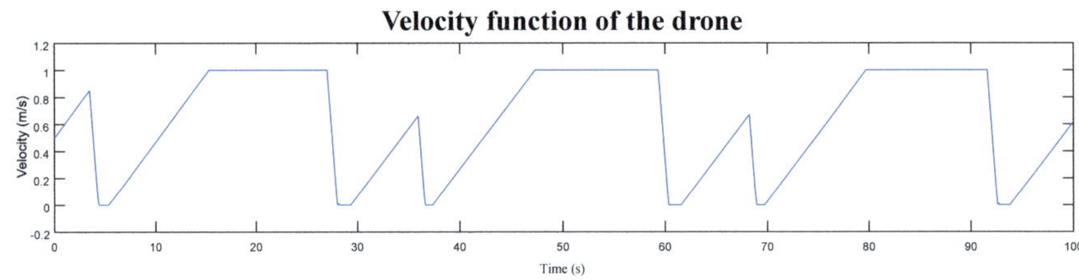

(a) Velocity of the drone

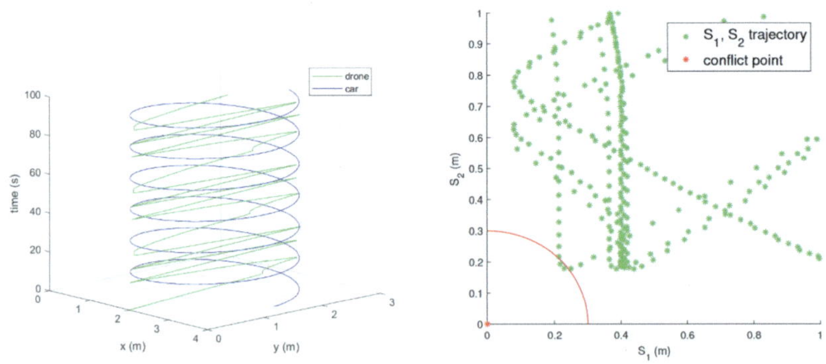

(b) Drone and vehicle trajectories (c) Illustration on $S_1 - S_2$ space

Fig. 3 Drone motion in conflict situations

In conclusion, a way to detect conflict points has been successfully tested in the simulations. the solution can be upgraded with using of Mahalanobis distances [11], which determines the distance from a cluster of points and the shape of the cluster can be acknowledged. It has the additional benefit that the consideration of sphere-shaped objects is not necessary.

Acknowledgments The research was supported by the European Union within the framework of the National Laboratory for Autonomous Systems (RRF-2.3.1-21-2022-00002). The paper was partially funded by the National Research, Development, and Innovation Office (NKFIH) under OTKA Grant Agreement No. K 135512.

References

1. Aggarwal S, Kumar N (2020) Path planning techniques for unmanned aerial vehicles: a review, solutions, and challenges. Computer Comm 149:270–299
2. Quan L, Han L, Zhou B, Shen S, Gao F (2020) Survey of UAV motion planning. IET Cyber-Syst Robot 2(1):14–21
3. Elmokadem T, Savkin AV (2021) A hybrid approach for autonomous collision-free UAV navigation in 3D partially unknown dynamic environments. Drones 5(3)
4. Chen Y-B, Luo G-C, Mei Y-S, Yu J-Q, Su X-L (2016) UAV path planning using artificial potential field method updated by optimal control theory. Int J Syst Sci 47(6):1407–1420
5. Tsiu L, Markus ED (2020) A survey of formation control for multiple mobile robotic systems. Int J Mech Eng Robot Res 9:1515–1520
6. Swadi SM, Kadhim AK, Ali GM (2023) Design of path planning controller of autonomous wheeled mobile robot based on triple pendulum behaviour. Int J Mech Eng Robot Res 12(1):23–31
7. Hegedu″s T, N´emethB, Ga´spa´r P (2019) Graph-based multi-vehicle overtaking strategy for autonomous vehicles. IFAC-PapersOnLine 52(5):372–377. In: 9th IFAC Symposium on advances in automotive control AAC 2019
8. N´emeth B, Hegedu″s T, G´aspa´r P (2018) Optimal control of overtaking maneuver for intelligent vehicles. J Adv Transp 2195760
9. Tompos D, N´emeth B (2023) Safe trajectory design for indoor drones using reinforcement-learning-based methods. In: 2023 IEEE 17th international symposium on applied computational intelligence and informatics (SACI), pp 27–32
10. Matlab optimization toolbox, R2023a, the MathWorks, Natick, MA, USA
11. McLachlan GJ (1999) Mahalanobis distance. Resonance 4(6):20–26

PID Controller Design for an Active Air Suspension System on Passenger Bus

Trong Tu Do, Van Tan Vu, Olivier Sename, and Peter Gaspar

1 Introduction

When a bus moves, its wheels always stay in contact with the road, but if the axle rigidly connects to the vehicle body, it can lead to the vehicle copying the road surface, causing dynamic loads that reduce passenger comfort and traffic safety. To mitigate these issues, vehicles use a suspension system to connect the chassis and wheels softly, isolating external vibrations for enhanced driver and passenger comfort and improved safety [1]. However, conventional leaf spring suspensions with unchanging stiffness result in a rough ride, limited shock absorption, and drawbacks like reduced off-road performance and increased weight, contributing to inconsistent handling and reduced traction [2, 3].

Recent advancements in materials technology and mechatronics have led to the developing of a cutting-edge suspension system known as the air suspension system [4, 5]. They provide adjustable ride height, allowing vehicles to adapt to varying road

T. T. Do
Faculty of Mechanical-Automotive and Civil Engineering, Electric Power University, Hanoi, Vietnam
e-mail: tudt@epu.edu.vn

V. T. Vu (✉)
Department of Automotive Mechanical Engineering, Faculty of Mechanical Engineering, University of Transport and Communications, Hanoi, Vietnam
e-mail: vvtan@utc.edu.vn

O. Sename
University Grenoble Alpes, CNRS, Grenoble INP, GIPSA-Lab, Grenoble, France
e-mail: olivier.sename@grenoble-inp.fr

P. Gaspar
Systems and Control Laboratory, Institute for Computer Science and Control (SZTAKI), Hungarian Research Network (HUN-REN), Budapest, Hungary
e-mail: gaspar@sztaki.hun-ren.hu

G. L. Conte and O. Sename (eds.), *Proceedings of 12th International Conference on Mechatronics and Control Engineering*, Lecture Notes in Mechanical Engineering,
https://doi.org/10.1007/978-981-96-6452-8_8

conditions or specific driving needs [6, 7]. This feature is particularly valuable for off-road vehicles and those requiring load leveling to maintain consistent handling. Moreover, air suspension enhances ride comfort by offering a smooth and cushioned experience, minimizing the impact of road imperfections [8]. The ability to fine-tune damping settings further refines the ride quality, catering to driver preferences and diverse driving scenarios by interaction with Proportional–Integral–Derivative (PID) [9, 10]. Fuzzy and its combination with PID controllers are designed in [11, 12] to intuitively and accurately reflect the driving performance with air suspension on buses and heavy trucks. In the research [13], the authors used particle swarm optimization turns LQR's weighting parameters for an adaptive air suspension system to enhance ride comfort. A genetic algorithm was constructed in [14] to optimize the parameters of the air spring element in a bus with 15 degrees of freedom. Additionally, air suspension systems contribute to improved handling, stability, and traction, while their durability and reduced maintenance requirements make them an appealing choice for long-term vehicle ownership [4, 15]. Beyond functionality, air suspension can also be utilized for aesthetic purposes, enabling vehicles to lower their stance for a sporty look when parked and raise it for practicality when on the move [16]. Central to the functioning of this system is the pneumatic spring, which serves as the pivotal elastic component. Pneumatic springs operate based on the compressibility of air, allowing them to adapt automatically and continuously to the prevailing driving conditions [17, 18].

Within the framework of this empirical investigation, the study delves into a fundamental aspect of vehicular dynamics, namely, the impact of leaf spring stiffness variation on suspension performance. This research explores the dynamic consequences of altering leaf spring stiffness and proposes a novel suspension paradigm for passenger vehicles. The proposed paradigm entails the substitution of traditional leaf springs with air suspension system, representing a paradigm shift in automotive suspension design. Within this novel framework, the air suspension system incorporates a PID controller that adjusts the air spring pressure, thus orchestrating a finely tuned balance in the vehicle's response to road irregularities. The core objective is to mitigate the oscillatory characteristics of the sprung mass's acceleration, ultimately leading to a discernible augmentation in the vehicle's ride comfort—an essential facet of the passenger experience.

2 Air Suspension System in Quarter Passenger Bus Model

2.1 Air Spring Model

The mechanical behavior of an air spring is intricately tied to dynamic and thermodynamic processes within the compressed air system. It can be mathematically modeled using the physical model shown in Fig. 1a, serving as a foundation for exploring and analyzing the pneumatic spring's mechanical characteristics. The airflow between the air spring and the auxiliary air tank is characterized by the mass m_m and the displacement w_m, and z_a is the vertical displacement of the air spring.

Figure 1b presents the mathematical model of the vertical load F_z acting on the air spring.

$$F_z = (p_{bo} - p_a)A_e + k_e z_s + k_v(z_s - w_s) \tag{1}$$

The conversion displacement of the volume of compressed air inside the pipeline w_s is calculated in (2).

$$w_s = \frac{A_{s(t)}}{A_e}\frac{V_{bo} + V_{ro}}{V_{ro}} w_m \tag{2}$$

in which k_e is a component that characterizes the hardness of compressed air in an air spring $k_e = \frac{p_a A_e^2 n}{V_{bo}+V_{ro}}$, k_v is a component that characterizes the hardness of compressed air in the accumulator $k_v = \frac{p_a A_e^2 n}{V_{bo}+V_{ro}}\left(\frac{V_{ro}}{V_{bo}}\right)$, M is the converted volume of compressed air in the pipeline, and c_w is the converted resistance coefficient of compressed air moving in the pipeline. The dynamics of the compressed air mass moves in the pipe can be described by

$$M\ddot{w}_s = k_v(z_s - w_s) - c_w|\dot{z}_s - \dot{w}_s|^{\beta} sign(\dot{z}_s - \dot{w}_s) \tag{3}$$

From (3), it is clearly seen that the cross-sectional area of the tube connecting the air spring and the accumulator simultaneously affects the entire block's converted amount of compressed air M and the converted resistance coefficient of compressed air moving in the pipeline c_w. Figure 2 presents the relationship between the cross-sectional area $A_{s(t)}$ and the equivalent drag coefficient of the air compression moves in the pipeline and the relationship between the cross-sectional area and the equivalent mass of the air compression moves in the pipeline, respectively.

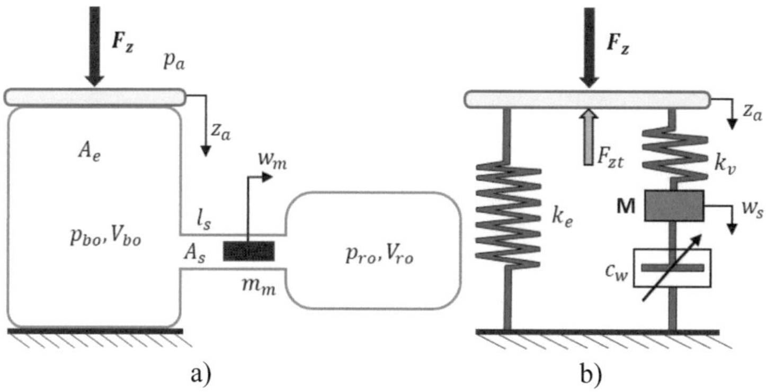

Fig. 1 Physical model of air spring (**a**) and its mathematical model (**b**) [16]

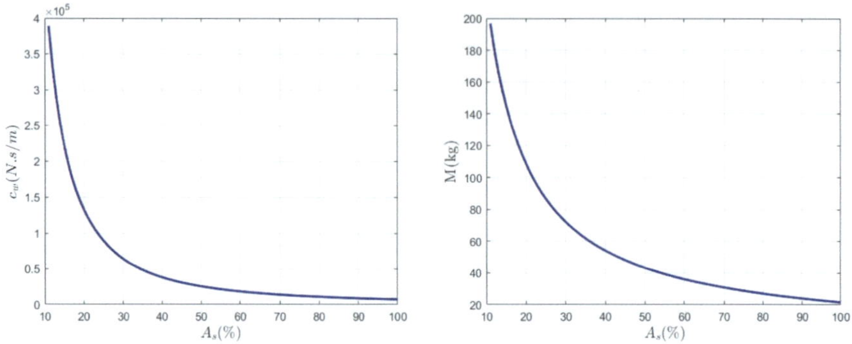

a) b)

Fig. 2 Relationship between the cross-sectional area and the equivalent drag coefficient. **a** The equivalent mass **b** of the air compression moves in the pipeline

2.2 Integrate the Air Spring into the Suspension System on Quarter Bus

Figure 3 is a quarter model of a passenger bus with two degrees of freedom. While Fig. 3a presents the leaf suspension system by a damper with a damping coefficient c_s and the leaf spring stiffness k_s, Fig. 3b illustrates a controlled air suspension system with an elastic element presented in Sect. 2.1. Their parameters are presented in Table 1. Considering that the suspension system in the quarter-passenger bus model is stable around the equilibrium, the system's vibration equation for the leaf suspension system in (4) and the air suspension system in (5) is

$$\begin{cases} m_s\ddot{z}_s + c_s(z_s - z_u) + k_s(\dot{z}_s - \dot{z}_u) = 0 \\ m_u\ddot{z}_u - c_s(z_s - z_u) - k_s(\dot{z}_s - \dot{z}_u) + k_t(z_u - q) = 0 \end{cases} \tag{4}$$

$$\begin{cases} m_s\ddot{z}_s + c_s(z_s - z_u) + k_e(\dot{z}_s - \dot{z}_u) + k_v(z_s - w_s) = 0 \\ m_u\ddot{z}_u - c_s(z_s - z_u) - k_e(\dot{z}_s - \dot{z}_u) - c_w|\dot{z}_s - \dot{w}_s|^\beta sign(\dot{z}_s - \dot{w}_s) + k_t(z_u - q) = 0 \end{cases} \tag{5}$$

From Eqs. (3) and (5), the dynamical equations of the quarter-passenger bus model using the air suspension system are written in Eq. (6). The parameter of the model is found in [16].

$$\begin{cases} m_s\ddot{z}_s + c_s(z_s - z_u) + k_e(\dot{z}_s - \dot{z}_u) + k_v(z_s - w_s) = 0 \\ m_u\ddot{z}_u - c_s(z_s - z_u) - k_e(\dot{z}_s - \dot{z}_u) - c_w|\dot{z}_s - \dot{w}_s|^\beta sign(\dot{z}_s - \dot{w}_s) + k_t(z_u - q) = 0 \\ M\ddot{w}_s - k_v(z_s - w_s) + c_w|\dot{z}_s - \dot{w}_s|^\beta sign(\dot{z}_s - \dot{w}_s) = 0 \end{cases} \tag{6}$$

3 Design PID Controller for Air Suspension System

A PID controller is known as the SISO control approach and is extensively employed across diverse engineering and industrial domains. It is made of three main actions: proportional, integral, and derivative. The global objective is to ensure closed-loop stability together with tracking performances (the observed process variable should follow a desired setpoint). The detailed scheme of the proposed PID controller in this research is illustrated in Fig. 4.

$$A_{s(t)} = K_p \ddot{z}_s(t) + K_i \int_0^t \ddot{z}_s(\tau)d\tau + K_d \frac{d\ddot{z}_s(t)}{dt} \tag{7}$$

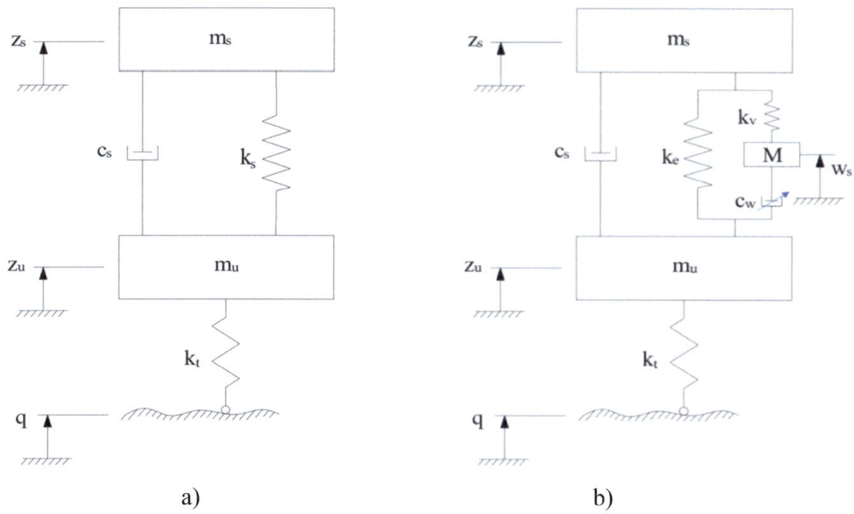

Fig. 3 Quarter bus model with **a** Leaf suspension and **b** Air suspension system [16]

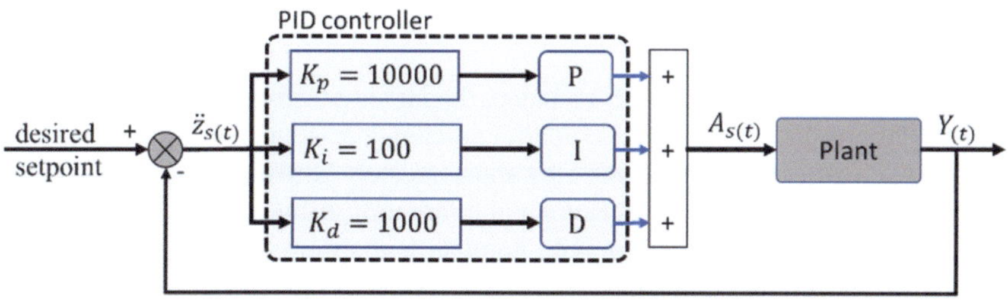

Fig. 4 Principle PID controller for controlling air suspension system

Table 1 Comparison reduction of r.m.s variables in frequency ranges

r.m.s	7÷10 rad/s		60÷80 rad/s	
	$\ddot{z}_s$	z_s	$\ddot{z}_s$	z_s
k_s maximum	–	–	–	–
k_s nominal	35.17%	−13.11%	−21.34%	49.35%
k_s minimum	51.70%	−8.52%	16.60%	74.15%
PID controller	**83.76%**	**63.28%**	**18.22%**	**76.14%**

Intending to increase comfort and stability for passenger bus, this study uses the sprung mass acceleration as the error to the PID controller to modify the cross-sectional area of air in the pipeline to adjust the stiffness of the air spring as given in Eq. (7). Selecting these coefficients normally by PID tuning methods, control algorithms, and advanced variations, which depend on designers' abilities to enhance their capabilities and address complex control challenges. In this research, the K_p, K_i, and K_d parameters are selected as 10000, 100, and 1000, respectively. Human vibrations experienced in vehicles are assessed across three levels: comfort, fatigue (permitting fluctuations within acceptable intensity levels), and the limit level (constrained based on the impact of fluctuations on health) [19]. Evaluation is conducted using the root mean square (r.m.s) index, as outlined in [16].

4 Simulation Results and Analysis

4.1 Simulation Results in Frequency Domain

In the Fig. 5, the correlation between the r.m.s of sprung mass acceleration and sprung mass displacement signals during sinewave excitation (0–100 rad/s) is depicted. Comparative assessment values for various stiffness levels in the leaf spring model are juxtaposed with the air spring suspension system featuring an integrated PID controller. The results show that as the stiffness of the leaf spring increases, the r.m.s value of the system equipped with leaf springs increases as its stiffness increases, and vice versa.

In the low-frequency range (7–10 rad/s), the air suspension system with an integrated PID controller demonstrates superior efficiency. It reduces the r.m.s of acceleration and velocity of the sprung mass by approximately 28% and 30%, respectively, compared to the leaf spring suspension with minimum stiffness. Furthermore, when compared to a suspension system with nominal leaf spring stiffness, the proposed model exhibits a more significant decrease in amplitude, with reductions of around 30 and 40% in acceleration and velocity of the sprung mass, corresponding to displacement and acceleration. In the high-frequency range (60–80 rad/s), air spring suspension signals have significantly lower r.m.s magnitudes than spring-based models,

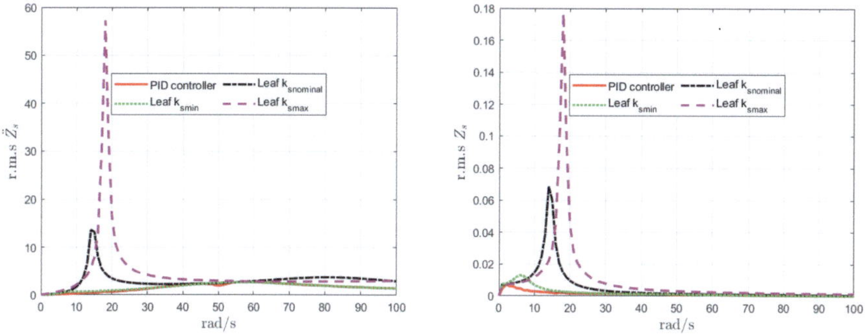

a) b)

Fig. 5 Frequency response of sprung mass acceleration (**a**), sprung mass displacement (**b**)

including leaf springs with varying stiffness coefficients (nominal, minimum, and maximum). Analyzing mass displacement, the highest leaf spring stiffness ranges from 2.5 to 1.6 mm, the nominal leaf spring from 1.1 to 0.8 mm, and the lowest leaf spring from 0.76 to 0.34 mm. However, the air spring suspension model, equipped with a PID controller, shows considerably smaller r.m.s values for mass translation, fluctuating between 0.73 and 0.31 mm (Fig. 7).

4.2 Simulation Results in Time Domain

In this study, the proposed models' characteristics are evaluated under driving scenarios on road ISO 8608-2016 class C at 60 km/h [20].

Figures 6–8 show simulation results comparing the leaf spring suspension system with varying spring stiffness to the air suspension system integrated with a PID controller. The dash green, dash blue, and dash black lines represent the leaf spring system with minimum, maximum, and nominal stiffness, respectively, marked by "o," "x," and no marker, while the solid red line depicts the simulation results of the air suspension system with the proposed PID controller.

Simulation results show a significant decrease in signals when using an air suspension system using the proposed PID controller. The displacement and acceleration signals of the sprung mass in Fig. 6 are reduced by about 34.2% and 87.1%. Therefore, the proposed air suspension system model proves to be highly effective in optimizing ride comfort criteria for the quarter-passenger bus.

As stiffness increases, the spring's generated force also increases, as depicted in Fig. 8a. In such instances, the pneumatic spring suspension model employs a PID controller to regulate pressure, effectively utilizing the spring force at its minimal level that fits the actuator's requirements [16]. Consequently, this approach significantly reduces the dynamic force transferred through the wheels to the road surface

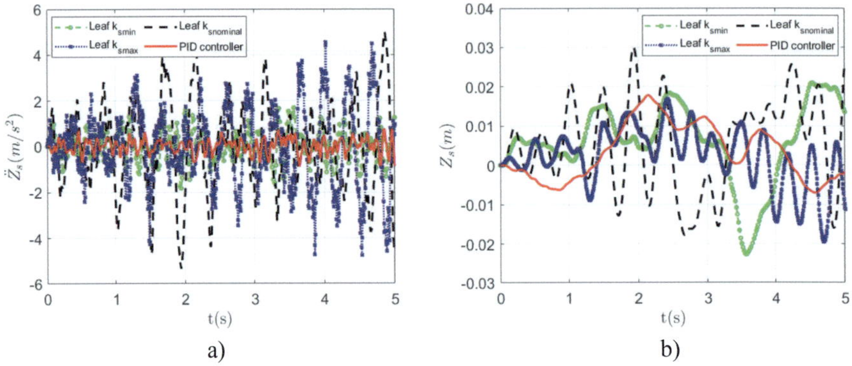

Fig. 6 Time response of sprung mass acceleration (**a**) Displacement (**b**) in the vertical direction of the passenger bus with road class C at 60km/h

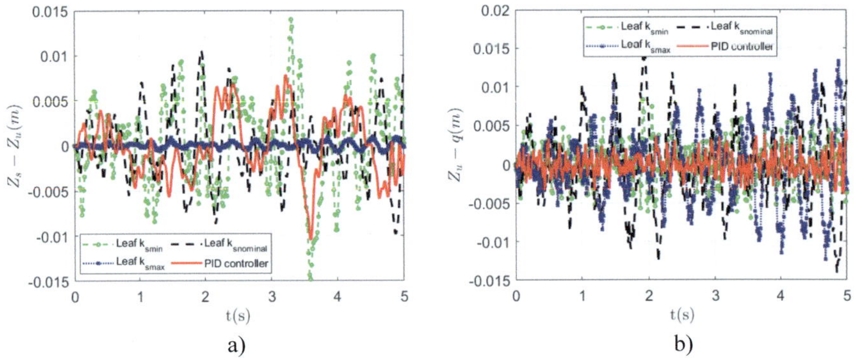

Fig. 7 Time response of $z_s - z_u$. **a** and $z_u - q$ **b** in the vertical direction of the passenger bus with road class C at 60 km/h

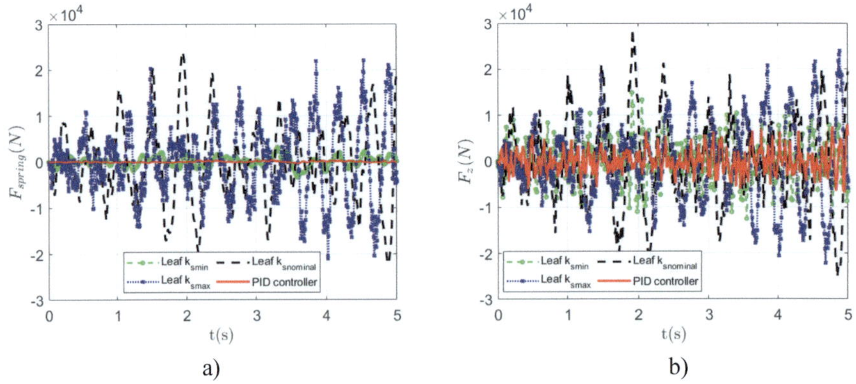

Fig. 8 Time response of spring force. **a** Tire forces. **b** In the vertical direction of the passenger bus with road class C at 60 km/h

Table 2 The values of the r.m.s of the signals

Signals	$\ddot{z}_s$	z_s	$z_s - z_u$	$z_u - q$
k_s minimum	0.684	0.009	0.0049	0.0028
k_s nominal	1.867	0.010	0.0038	0.0053
k_s maximum	1.573	**0.007**	**0.0003**	0.0047
PID controller	**0.242**	**0.007**	0.0035	**0.0017**

(Fig. 8b). When the suspension system utilizes leaf springs with the highest stiffness, the impact force on the road surface reaches approximately 3.23 kN. However, within the proposed model, this value is notably reduced to approximately 1.61 kN, resulting in a roughly 50% reduction in the vehicle's dynamic force on the road surface.

Table 2 displays the r.m.s statistics for signals related to the sprung mass fraction in the quarter model suspension system of a passenger bus. In leaf spring suspension models, r.m.s values of sprung mass acceleration and displacement are notably higher than those in the air suspension system. Specifically, the leaf spring's minimum stiffness yields the most significant acceleration 0.684 m/s^2. In contrast, the air suspension system with the proposed PID controller consistently exhibits the smallest values in both road surface profile excitations, with r.m.s values for sprung mass acceleration and displacement at 60 km/h on ISO standard road class C pavement measuring 0.242 m/s^2 and 0.007 m. Despite greater suspension displacement in the pneumatic system model, the relative displacement of the unsprung mass concerning the road surface is notably reduced to 0.0017 m.

5 Conclusion

In conclusion, this study explores a dynamic model for a quarter-passenger bus employing an air suspension system with adjustable stiffness via a PID controller. The proposed model enhances vehicle smoothness, reducing sprung mass acceleration by 76.6% and displacement by approximately 34.2% under ISO road surface excitations. Despite larger suspension displacement in the pneumatic spring system, dynamic force transmitted to the road is significantly reduced by about 50% compared to the nominal model. The pneumatic spring mechanism with PID controller achieves a remarkable force reduction of nearly 74% compared to leaf spring suspension with maximum stiffness, leading to low energy consumption. This research dawn on a balance between safety and comfort, contributing to the understanding of suspension system dynamics and advancing passenger vehicle ride quality. Future research will explore nonlinear characteristics of air spring pressure and employ optimal or robust control algorithms for multi-objective problems in vehicle air suspension systems.

References

1. Savaresi S, Poussot-Vassal C, Spelta C et al (2010) Semi-active suspension control design for vehicles
2. Varma N, Ahuja R, Vijayakumar T, Kannan C (2021) Design and analysis of composite mono leaf spring for passenger cars. Mater Today Proc 46:7090–7098. https://doi.org/10.1016/j.matpr.2020.10.073
3. Shohel SM, Sen GS, Riyad SH (2022) Weight optimization and finite element analysis of automobile leaf spring: a design construction referred to electric vehicle. IOP Conf Ser Mater Sci Eng 1259:012024. https://doi.org/10.1088/1757-899x/1259/1/012024
4. Pradhan P, Singh D (2021) Review on air suspension system. Mater Today Proc 81:486–488. https://doi.org/10.1016/j.matpr.2021.03.640
5. Karimi Eskandary P, Khajepour A, A. Wong MA (2016) Analysis and optimization of air suspension system with independent height and stiffness tuning. Int J Automot Technol 17:807–816. https://doi.org/10.1007/s12239-016-0079-9
6. Ma X, Wong PK, Zhao J et al (2018) Design and testing of a nonlinear model predictive controller for ride height control of automotive semi-active air suspension systems. IEEE Access 6:63777–63793. https://doi.org/10.1109/ACCESS.2018.2876496
7. Zhao R, Xie W, Zhao J et al (2021) Nonlinear ride height control of active air suspension system with output constraints and time-varying disturbances. Sensors 21:1–16. https://doi.org/10.3390/s21041539
8. Chen G, Jiang Y, Tang Y, Xu X (2023) Revised adaptive active disturbance rejection sliding mode control strategy for vertical stability of active hydro-pneumatic suspension. ISA Trans 132:490–507. https://doi.org/10.1016/j.isatra.2022.06.008
9. Jiang X, Cheng T (2023) Design of a BP neural network PID controller for an air suspension system by considering the stiffness of rubber bellows. Alexandria Eng J 74:65–78. https://doi.org/10.1016/j.aej.2023.05.012
10. Zhang P, Chen WW, Wang H, et al (2022) Fuzzy PID control method for damping of electronically controlled air suspension shock absorbers for vehicles. Eng Res Express 4. https://doi.org/10.1088/2631-8695/aca13a
11. Xia Q, Bai R, Wang H (2018) Fuzzy control of damping force in the air suspension system. In: 2018 international conference on information, cybernetics, and computational social systems ICCSS, pp 328–330. https://doi.org/10.1109/ICCSS.2018.8572384
12. Bai X, Lu L, Zhang C, Geng W (2023) Research on height adjustment characteristics of heavy vehicle active air suspension based on fuzzy control. World Electr Veh J 14. https://doi.org/10.3390/wevj14080210
13. Gp S, Mm K (2019) A contemporary adaptive air suspension using LQR control for passenger vehicles. ISA Trans 93:244–254. https://doi.org/10.1016/j.isatra.2019.02.031
14. Hung TM (2022) Optimal selection for an air suspension system on buses through a unique high level parameter in genetic algorithms. Heliyon 8:e09059. https://doi.org/10.1016/j.heliyon.2022.e09059
15. Nazemian H, Masih-Tehrani M (2020) Development of an optimized game controller for energy saving in a novel interconnected air suspension system. Proc Inst Mech Eng Part D J Automob Eng 234:3068–3080. https://doi.org/10.1177/0954407020927147
16. Vu VT, Hung TM, Sename O (2021) An investigation into the ride comfort of buses using an air suspension system. Int J Heavy Veh Syst 28:184–205. https://doi.org/10.1504/IJHVS.2021.115595
17. Le VQ (2017) Comparing the performance of suspension system of semi-trailer truck with two air suspension systems. Vibroengineering Procedia 14:220–226. https://doi.org/10.21595/vp.2017.19224
18. Tuan N Van, Quynh L Van, Thao VTP, Duy LQ (2020) Optimal design parameters of air suspension systems for semi-trailer truck. Part 2: Results and discussion. Vibroengineering Procedia 33:147–152. https://doi.org/10.21595/vp.2020.21563

19. International Standard Orgainization (2018) ISO 2631-5: Mechanical vibration and shock—evaluation of human exposure to whole-body vibration
20. International Standard Orgainization (2016) ISO 8608:2016 Mechanical vibration—road surface profiles—reporting of measured data

Theoretical Investigation and Simulation of a Brake-by-Wire Based Bicycle Wedge Brake

Bence Nagy and Dénes Fodor

1 Introduction

In recent years high efforts are made to make road traffic safer. For the automotive industry many standards were introduced to achieve functional safety of the produced vehicles, like ISO 26262 or SOTIF, and for motorbikes and bicycles, many new equipment appeared on the market in order to ensure rider safety.

However, bicycles are much simpler constructions than the motorized vehicles, thus smaller focus is drawn to ensure the safety of bicycles. In the last years the usage of bicycles and e-bikes is increasing and high number of accidents were suffered by bicycle drivers [2]. In many cases, cyclists don't wear essential protection, so a minor accident can cause major injuries.

With the appearance and broad usage of electric bicycles, the possibility of implementing electric safety systems in e-bikes is becoming more and more feasible. With this in mind, we can conclude that the use of anti-lock brake systems is feasible and can increase rider safety. Commercial anti-lock brake systems started to appear on the market and are becoming popular among e-bike users. Those brake systems are mainly hydraulic, and thus require high maintenance and complicated control systems [1].

In this paper we will focus on evaluating an electro-mechanic ABS system that uses a mechanic wedge brake system instead of a hydraulic brake system. The main objective of this research paper is to carry out a theoretical investigation of an electrically actuated bicycle anti-lock brake system. We will derive the mathematical model of a bicycle system with a mechanical wedge brake actuated by a DC motor. Using

B. Nagy (✉) · D. Fodor
Department of Power Electronics and E-Drives, Széchenyi István University, Győr, Hungary
e-mail: bence.nagy147@gmail.com

D. Fodor
Vehicle Industry Research Center, Széchenyi István University, Győr, Hungary

G. L. Conte and O. Sename (eds.), *Proceedings of 12th International Conference on Mechatronics and Control Engineering*, Lecture Notes in Mechanical Engineering,
https://doi.org/10.1007/978-981-96-6452-8_9

this model we will prove the working principle of the wedge brake ABS system and we will use this proof of concept as a basis for further research projects.

The main advantage of a wedge brake-based ABS system is that it requires less maintenance and is also cheaper to produce as the main parts are mechanic and doesn't require hydraulic components. However, the control of such systems has to be put in focus in order to receive an efficient ABS for e-bikes [1].

2 Mathematical Model of the Wedge Brake System

This research paper captures the beginning of an electro-mechanic wedge brake ABS development for electric bicycles. As we are evaluating different design- and control concepts, it is crucial to use virtual prototypes in order to reduce the need of physical prototypes. By using the process of Model-Based Design, we can not only reduce the dependence on physical prototypes, but we also gain efficiency in the design process [3]. In Model-Based Design, the development process centers around a system model—this model is used to capture the plant dynamics (in our case the dynamic model of a bicycle and wedge brake model) and the control model. From this description we can conclude that using the Model-Based Design methodology, we can fulfill the research objective of investigating the working concept of the bicycle wedge brake system, and with deriving the mathematical models for the main systems we can also avoid the need of physical prototypes.

Using the Model-Based Design for the wedge brake system, we can brake down the system into four main subsystems:

1. Wedge brake model
2. Electrical actuation model
3. Longitudinal bicycle model
4. Anti-lock brake control model.

2.1 *Wedge Brake Model*

The main concept of a wedge brake is described in Fig. 1. The brake lining is equipped with a small wedge on one of its sides. The wedge is pressed by an external force (from the actuator) and thus a pressure will be generated on the braking disc. The braking force will be generated from the contact between the brake lining and the brake disc. If we draw the mechanical model of the wedge brake, we get the mechanical system shown in Fig. 2. Figure 2 shows the forces acting on the wedge brake during braking. We can express the relationship between the motor force (F_m) and the braking

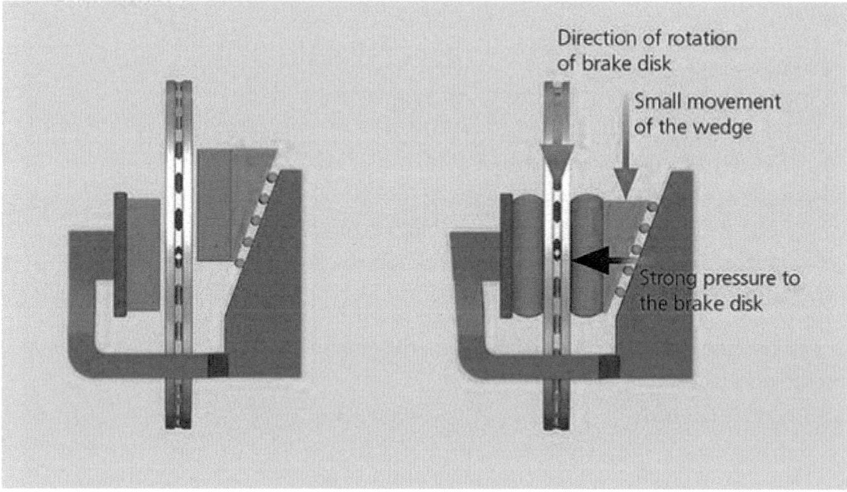

Fig. 1 Operating concept of a wedge brake

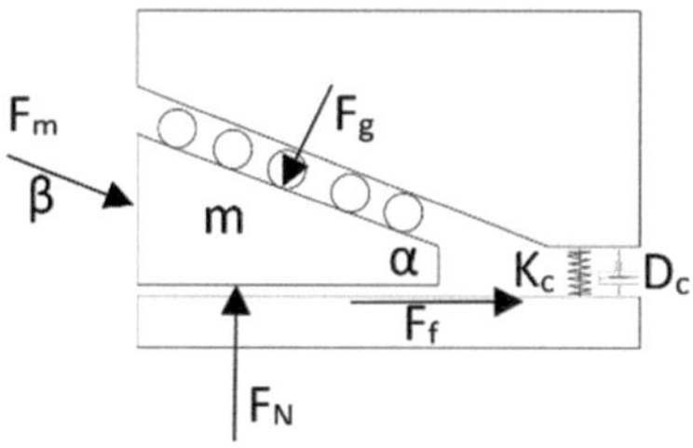

Fig. 2 Mechanical model of a wedge brake

force (F_f) using the angle of the wedge (α) and the friction coefficient (μ) between the wedge and the brake disc.

$$F_m = F_f \frac{\tan\alpha - \mu}{\mu}$$

We can also notice that the brake disc and the brake lining are connected with a spring and damper connection representing the brake pad in the system.

Based on the free body diagram shown in Fig. 2 we can derive the equations of motion of the wedge brake based on the second law of motion. The equations in each direction will look as follows:

$$ma_x = F_m \cos\beta + F_f - F_g \sin\alpha \tag{1}$$

$$ma_y = F_m \sin\beta + F_N + F_g \cos\alpha \tag{2}$$

where:

- m: mass of the wedge
- a_x: acceleration of the wedge in x direction
- a_y: acceleration of the wedge in y direction
- F_m: motor force acting on the wedge
- F_f: braking force
- F_g: force between the brake lining and the wedge
- F_N: normal force between the brake disc and the wedge
- α: inclination angle of the wedge

The braking force can be expressed as the function of the friction coefficient μ and the normal force F_N:

$$F_f = F_N \mu \tag{3}$$

In the current status we have two equations of motion, one for x and one for y direction. For our use case it is crucial to express the displacement in the x direction and find a relation between the two directions. If we consider the geometrical characteristics of the wedge we can derive the following function between x and y coordinates:

$$y = \tan \alpha x \tag{4}$$

After substituting Eqs. 3, 4 to the Eqs. 1 and 2 we end up with the following equations:

$$m\ddot{x} = F_m \cos \beta + \mu F_N - F_g \sin \alpha \tag{5}$$

$$m\ddot{x} \tan \alpha = F_m \sin \beta + F_N + F_g \cos \alpha \tag{6}$$

We can see that both Eqs. 5 and 6 contain F_g. We can use this characteristics to combine the two equations and express the acceleration in the x direction. Furthermore, from the characteristics of the wedge and the design decisions, we can find a characteristics where the angle of the motor force β is close to the wedge angle α. In that case we can assume that:

$$\beta \approx \alpha \tag{7}$$

After combining Eqs. 5 and 6 we end up with:

$$m\ddot{x}(1 + \tan^2 \alpha) = F_m(\sin \alpha \tan \alpha + \cos \alpha) + F_N(\mu - \tan \alpha) \tag{8}$$

The force F_N acting between the brake disk and the wedge can be expressed using the spring and damper from the mechanical model:

$$F_N = \tan \alpha \,(K_c x + D_c \dot{x}) \tag{9}$$

After substituting Eqs. 9–2.1 we get the final equation of the acceleration in x direction:

$$\ddot{x} = \frac{1}{m\left(1+\tan^2\alpha\right)}\left(\frac{F_m}{\cos\alpha} + \tan\alpha\,(K_c x + D_c \dot{x})\,(\mu - \tan\alpha)\right) \tag{10}$$

2.2 Longitudinal Model of a Bicycle

In order to test the wedge brake model virtually, we need to obtain a longitudinal model for the dynamic characteristics of the bicycle. As we are at the beginning of the development process, a low-fidelity model is sufficient for this case. The only major requirement for the bicycle dynamics model is that it should take into account the road adhesion coefficient [4]. We will use the single wheel model for modeling the longitudinal dynamics of the bicycle. The model of the system is shown in Fig. 3.

It is important to note that we will control the front wheel using our ABS system. The importance of controlling the front wheel is due to the fact, that from stability point of view, a wheel lock is more crucial if the front wheel is involved than the rear one [5]. The velocity of the bicycle and the rotational speed of the wheel can be derived using the following equations:

$$m\dot{v} = F_x \tag{11}$$

$$J\dot{\omega} = -RF_x - M_f sign(\omega) \tag{12}$$

where:

- m: Mass of the whole bicycle system
- v: Longitudinal velocity of the bicycle

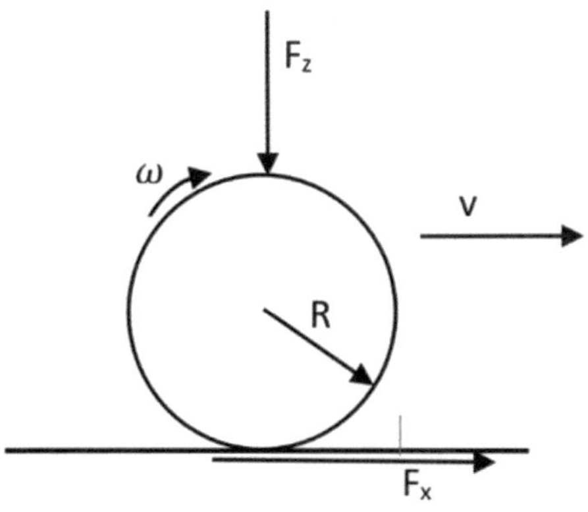

Fig. 3 Single wheel model with the acting forces

- F_x: Longitudinal force acting on the system
- J: Moment of inertia of the wheel
- ω: Rotational speed of the wheel
- R: Radius of the wheel
- M_f: Braking torque

The longitudinal force can be expressed as:

$$F_x = \mu F_z = \mu m g \tag{13}$$

Substituting Eqs. 13–Eqs. 11 and 12 we get:

$$m\dot{v} = \mu m g \tag{14}$$

$$J\dot{\omega} = -R\mu m g - M_f sign(\omega) \tag{15}$$

The friction coefficient is the function of the slip of the wheel (s) and different coefficients that can be expressed using the Magic Tire formula [7]:

$$\mu = D\sin\left(C\arctan\left(B\left((1-E)s - \frac{E}{B}\arctan Bs\right)\right)\right) \tag{16}$$

On dry road between concrete and tire material, the values of coefficients are the following:

- $D = 1$
- $C = 2.7$
- $B = 3.76$
- $E = 1.06$

The slip of the wheel can be expressed as the function of the longitudinal velocity v and the rotational speed of the wheel ω [7]:

$$s = \frac{R\omega - v}{max(R\omega, v)} \tag{17}$$

The denominator of Eq. 17 during acceleration and braking looks as follows:

$$acceleration : R\omega > v \tag{18}$$

$$braking : R\omega < v \tag{19}$$

As we are investigating the behavior of the system during braking, the Eq. 17 will be used in this format:

$$s = \frac{R\omega - v}{v} \tag{20}$$

2.3 DC Motor and Ball Screw Model

In the physical system the wedge brake is actuated using a DC motor and a ball screw, so in case we want to create a model of the real system, we need to implement also the DC motor model and a ball screw model.

DC motor model The DC motor model is shown in Fig. 4. The motor model is described using the differential equations:

$$\dot{\omega} = \frac{1}{J}(K_t i - b\omega) \tag{21}$$

$$\dot{i} = \frac{1}{L}(-Ri + U + K_e\omega) \tag{22}$$

where:

- ω: rotational speed of the motor
- J: motor moment of inertia
- K_t: torque constant of the motor
- i: electric current
- b: motor friction
- L: motor inductance
- R: motor resistance
- U: motor voltage
- K_e: electric force constant

Ball screw model The ball screw can be modeled using the equation:

$$M_m = F_m \frac{L}{2\pi} \frac{1}{0.6} \tag{23}$$

where:

- M_m: motor torque

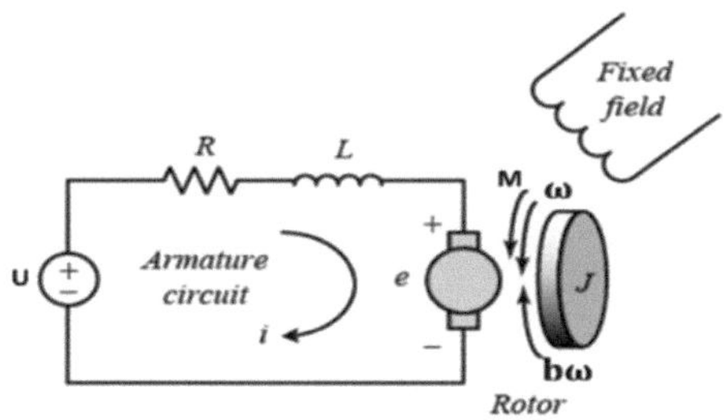

Fig. 4 DC motor model

- F_m: motor force
- L: ball screw thread pitch

With the derived mathematical models of each major component of the wedge brake system we can now move on to applying the models in the model-based design workflow and continue the system analysis using simulations in Simulink.

3 Creating a Plant Model in Simulink

As we already have the derived mathematical models of the wedge brake system components we can implement the system in Simulink and start creating system-level analysis on the wedge brake system [6]. In order to maintain readability across the model, we will organize the model into 4 main subsystems:

1. DC Motor
2. Ball screw joint
3. Wedge brake model
4. Longitudinal single wheel model

The implemented system with the subsystems is shown in Fig. 5. We can clearly see the boundaries between each architectural unit, their inputs and outputs, and also the input and outputs of the whole system.

In Fig. 5 we can see the input and outputs of the system:

- Input:

  U: input voltage of the DC motor

- Outputs:

  v: longitudinal velocity of the wheel
  x: traveled distance of the wheel
  $omega$: rotational speed of the wheel
  sx: slip of the wheel

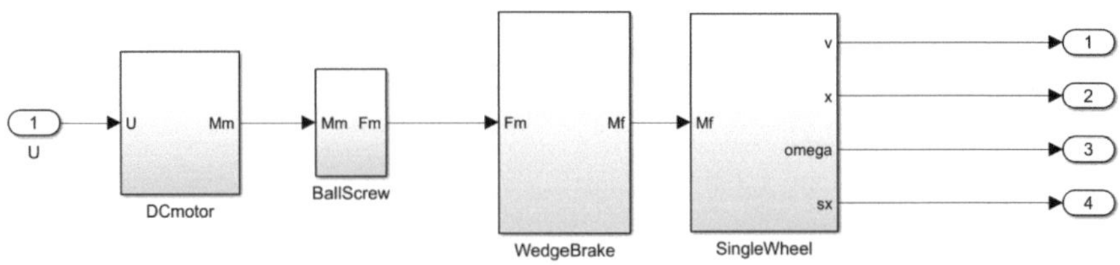

Fig. 5 Wedge brake system model with the DC motor, ball screw, wedge brake, and single wheel longitudinal model

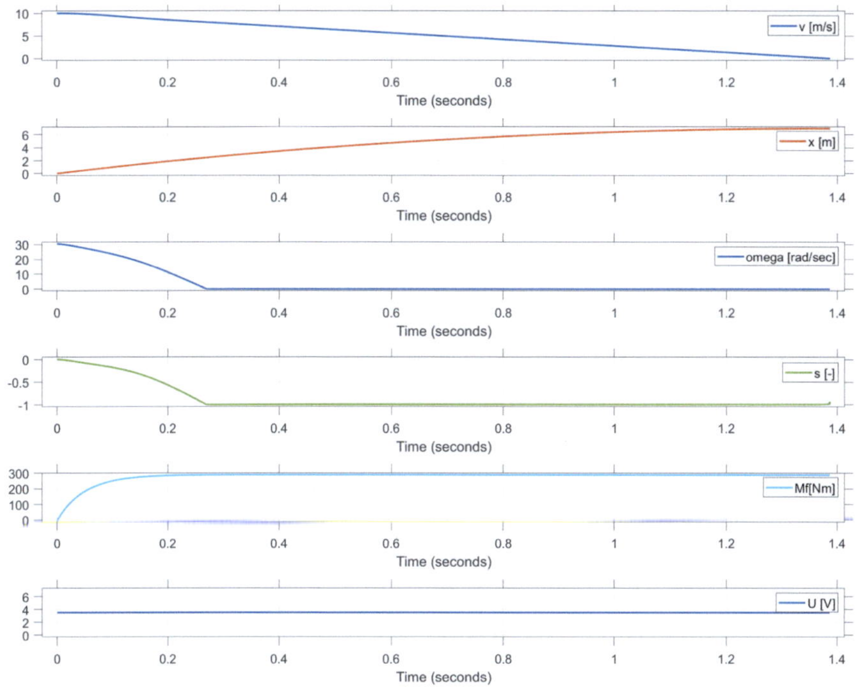

Fig. 6 Open-loop system reaction to $U = 3.5V$ DC motor input

In Fig. 5 we can also see that we feed the outputs of each subsystem as the input of the next subsystem, and we have no feedback loop implemented in the system, thus we are talking about an open-loop system. This open-loop implementation is sufficient to observe the system reaction to specific input voltages of the DC motor.

If we run the simulation with a constant $U = 3.5V$ voltage on the input we obtain the results shown in Fig. 6. From the simulation results we can see that we start braking at time $t = 0s$, and both the velocity v and the wheel rotational speed $omega$ start to decrease, but the rotational velocity rapidly drops to $omega = 0\ deg/s$ and we experience a wheel lock-up at the time between $t = 0.25 and 0.3s$. We can also see the wheel lock-up on the value of the wheel slip s, as it decreases to the value of -1 once the rotational speed is 0. This means that with a locked front wheel, the bicycle becomes instable and difficult to control, thus causing a danger to the driver. In order to avoid the lock-up of the wheel and increase the safety of the system, we need to introduce a control algorithm that controls the input voltage U of the DC motor.

4 Control Algorithm Implementation

In order to control the system we will introduce a control algorithm to control the input voltage U of the DC motor. For the control algorithm the bang-bang controller will be used. The working principle of the controller is shown in Fig. 7. The controller compares the actual slip value s to a reference slip value s_ref and can operate in two states:

1. $s_{ref} \leq s$: no control needed; $U = 3.5V$
2. $s_{ref} > s$: controller is active; $U = 0V$

The plant model with the implemented controller is shown in Fig. 8. We can observe that in Fig. 8 we use the slip s output of the SingleWheel Subsystem as the input of the BangBangControl Subsystem, and thus we receive a closed-loop model. After we run the simulation we obtain the results plotted in Fig. 9. Just as with the model without control, we start braking at the simulation time $t = 0s$ and drive $U = 3.5V$ voltage to the DC motor. However, with the implemented bang-bang controller the wheel slip s can not decrease below the reference value of s_{ref}, hence the control signal intervenes and reduces the input voltage to $U = 0V$. Thus the braking torque also decreases and the wheel does not lock, but the rotational speed ω of the wheel decreases seamlessly with the velocity of the bicycle v. With the implemented control we not only achieve a smooth decrease of the rotational speed of the wheel, but the braking distance is also reduced. Thus we not only create a more efficient braking system, but also increase the safety of the rider, as the bicycle remains stable under braking without wheel lock-up.

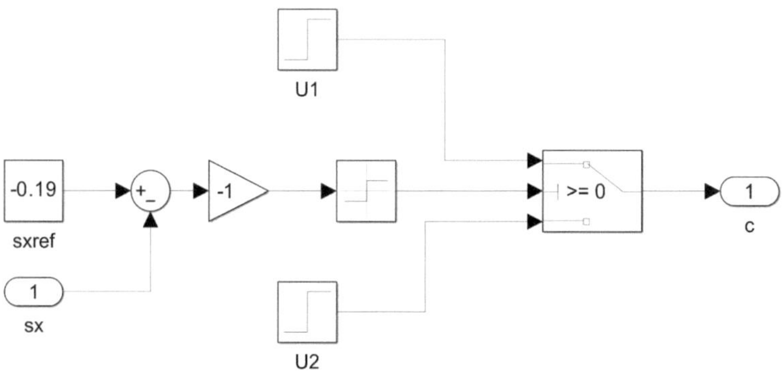

Fig. 7 Working concept of the bang-bang controller

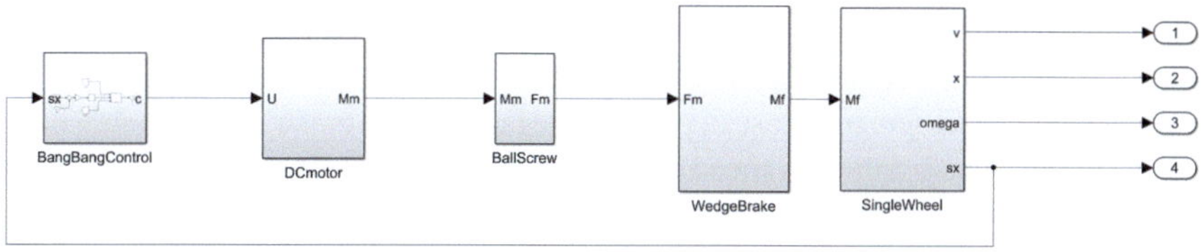

Fig. 8 Closed-loop model of the control algorithm and the plant model

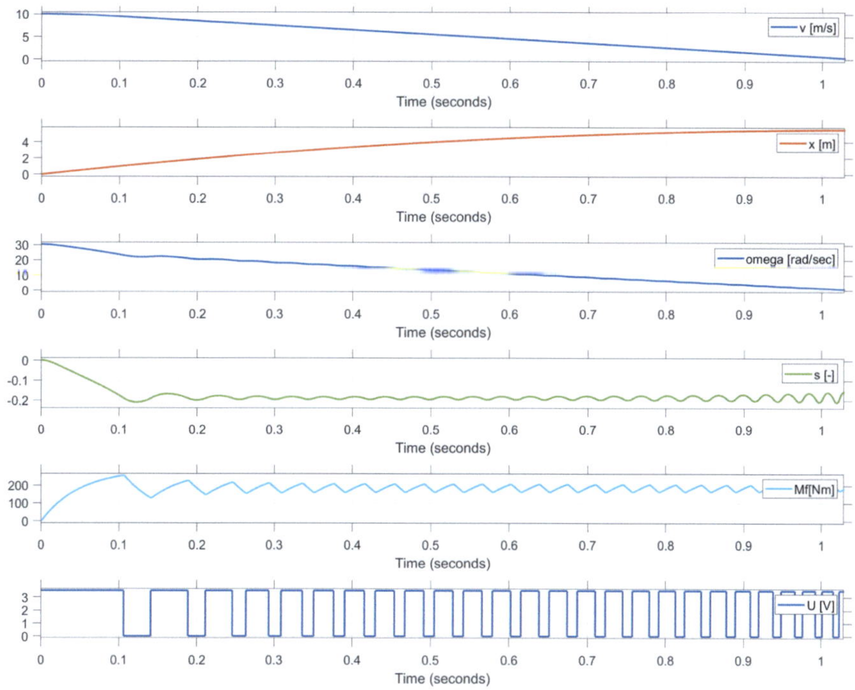

Fig. 9 System behavior with control algorithm

5 Conclusion

This paper aims to highlight the first phase of the development of an electro-mechanic bicycle wedge brake system. As many people use bicycles during their everyday life, it is important to prevent the bicycle riders from possible injuries. With this paper the aim was to highlight the working concept of a bicycle ABS system and to show the first phase of the wedge brake-based ABS system development. The ABS system described in this paper includes a novel mechanic wedge brake system, so in order to prove the working concept of the wedge brake, a simulation model was developed based on the mechanical behavior of the wedge brake system. Later, a bang-bang controller was implemented to model the feature of the ABS system.

From the results in Fig. 9 we see that the wedge brake system with the implemented control successfully brakes the system without locking the wheel of the bicycle.

In the next development phase, a prototype of the wedge brake mechanism will be developed and also a more advanced control algorithm may be implemented, in order to achieve a reliable operation of the wedge brake system during different conditions.

Acknowledgment I would like to acknowledge and give my thanks to my co-author, Dénes Fodor and the Vehicle Industry Research Center of the Széchenyi István University. The work was supported by Project no. TKP2021-NKTA-48 and has been implemented with the support provided by the Ministry of Technology and Industry of Hungary from the National Research, Development and Innovation Fund, financed under the TKP2021-NKTA funding scheme. The research was also supported by the European Union within the framework of the National Laboratory for Autonomous Systems (RRF-2.3.1-21-2022-00002).

References

1. Enisz K, Szalay I, Fodor D, Nagy K, Jakab R (2012) Bicycle anti-lock braking system prototype development acta universitatis sapientiae electrical and mechanical. Engineering 4:45–57
2. U.S. Department of Transportation, (2010), Traffic safety facts 2009: a compilation of motor vehicle crash data from the fatality analysis reporting system and the general estimates system, national highway traffic safety administration, Washington, DC 20590. http://www-nrd.nhtsa.dot.gov/Pubs/811402.pdf
3. Smith PF, Prabhu SM, Friedman JH Best practices for establishing a model-based design culture
4. Zhu Y, Li X, Zhang X, Li S, Liu Q, Yuan S (2022) Research on tire-road parameters estimation algorithm for skid-steered wheeled unmanned ground vehicle. Machines 10:1015. https://doi.org/10.3390/machines10111015
5. Corno M, D'Avico L, Savaresi SM (2018) An anti-lock braking system for bicycles. In: 2018 IEEE conference on control technology and applications (CCTA). https://doi.org/10.1109/CCTA.2018.8511615
6. Abd-Elhameed M, Emam A, El-Demerdash S, Shaban S, Mahmoud M (2011) A theoretical study of the self reinforcement brake system. J Am Sci
7. Pacejka HB (2012) Tire and vehicle dynamics. ISBN 978-0-08-097016-5. https://doi.org/10.1016/C2010-0-68548-8 Morgan Kaufmann, San Francisco (1999)

Symmetric Measurement Equations for Multi-target Tracking

Olivér Törö, Domonkos Csuzdi, László Lindenmaier, and Tamás Bécsi

1 Introduction

State estimation and tracking in a multi-target environment pose several difficulties. In real-world multi-target tracking (MTT) applications, we usually face issues such as background noise, clutter measurements, missed detections, or detections generated by more than one object. All of these issues are related to the data association problem; that is, we need to decide which object track should be updated with which measurement. A measurement reported by a sensor can originate from different sources. A measurement can be regarded as noise if generated by a random process, e.g., thermal noise inside the sensor or environmental effects exterior to the sensor. Clutter is a measurement generated by a real object that is not our target, and we do not intend to track its state. A measurement may have more than one source if the objects are too close to each other and cannot be resolved by the sensor. On the other hand, an object can be responsible for more than one measurement, which is usually the case for extended targets.

A data association procedure has to deal with all these issues to serve reliable information for a fusion algorithm. The simple and natural way to address the uncertainty in the data association is to try every possible variation and choose the best one according to some measure. This approach is feasible only for scenarios with a few targets and a low number of measurements. Given the simplest case, when the number of measurements is equal to the number of targets, there are no missed detections and no multiple detections of a target, we face a permutation problem. There are $N!$ ways to pair N measurements with N targets. If we have false detections besides the real ones, the first task is to select N from the total of $M > N$ measurements and then pair the selected ones with the N targets. The number of variations is $M!/(M-N)!$.

O. Törö (✉) · D. Csuzdi · L. Lindenmaier · T. Bécsi
Budapest University of Technology and Economics, Budapest, Hungary
e-mail: toro.oliver@kjk.bme.hu

G. L. Conte and O. Sename (eds.), *Proceedings of 12th International Conference on Mechatronics and Control Engineering*, Lecture Notes in Mechanical Engineering,
https://doi.org/10.1007/978-981-96-6452-8_10

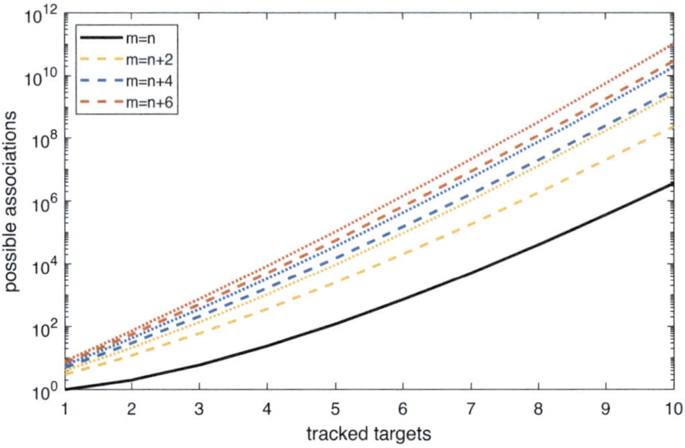

Fig. 1 Number of possible data association hypotheses as a function of the collected measurements M and the target number N. The dotted line represents the cases where miss detections can occur according to Eq. (1)

In scenarios when the probability of detecting a target is less than one, miss detections may occur, which makes the data association more difficult (Fig. 1). For example if we have two target tracks, x_1 and x_2, and two measurements, z_1 and z_2, the following associations are possible:

$$\begin{array}{lll} x_1 \leftrightarrow \varnothing, x_2 \leftrightarrow \varnothing & x_1 \leftrightarrow \varnothing, x_2 \leftrightarrow z_1 & x_1 \leftrightarrow z_1, x_2 \leftrightarrow \varnothing \\ x_1 \leftrightarrow z_1, x_2 \leftrightarrow z_2 & x_1 \leftrightarrow \varnothing, x_2 \leftrightarrow z_2 & x_1 \leftrightarrow z_2, x_2 \leftrightarrow \varnothing \\ x_1 \leftrightarrow z_2, x_2 \leftrightarrow z_1 & & \end{array}$$

Pairing an object with the empty set $\varnothing$ means that missed detection occurred and the measurement set does not contain a measurement originating from that target. In general, having M measurements and $N \leq M$ targets from which we detect $0 \leq n \leq N$ the number of possible associations is

$$\sum_{n=0}^{N} \binom{N}{n} \frac{M!}{(M-n)!}. \tag{1}$$

The combinatorial complexity of the data association problem does not allow us to reach the optimal solution via an exhaustive search; thus, certain approximate and heuristic methods are used in practice. The main goal is to reduce the number of possible association hypotheses to a computationally tractable amount. It is done on different abstraction levels. For a radar system, the constant false alarm rate (CFAR) detection is a typical method to estimate the threshold to filter out noise [12]. Clutter measurements are also undesirable; their modeling is not trivial and depends on the environment. Although modeling the clutter as a uniform Poisson point process is an option, additional information can be incorporated if the environment has specific characteristics, such as in road traffic situations [8].

In scenarios where the objects can be regarded as point-like, the standard measurement model is a usual assumption, which states that an object cannot generate more than one measurement and a measurement cannot be generated by more than one object. This assumption ensures that the measurement set should not be partitioned such that an object is paired with more than one measurement, meaning the association is one-to-one.

Regarding the assembly of feasible association hypotheses, several methods exist [16]. The simplest one is the global nearest neighbor filter, which selects the unique joint association of measurements with targets in a way that a cost function is minimized. The filtering update step applies the associated measurement for each target.

The joint probabilistic data association filter (JPDAF) considers multiple feasible associations and computes a weighted average of the estimations [2]. The weights are the marginal probabilities of certain associated measurement-track pairs. The JPDAF is a well-established method, used in practice efficiently [1]. Its downside is the combinatorial complexity and its consequence, the high peak-to-average ratio of computational requirement [7].

Another multi-target estimator family is the multiple hypotheses trackers (MHT) [16]. These algorithms maintain numerous hypotheses about the possible data associations, propagate concurrent possible object tracks, and defer the decision about the correct hypothesis until enough information is gathered. MHT trackers use more sophisticated association algorithms and can consider temporal correlations in the estimation process. The required computational power for these algorithms is even higher than that of the JPDAF.

A relatively new branch of MTT estimators is the random finite set (RFS) based algorithms [11]. The elements of an RFS are state vectors of objects being tracked. The RFS approach is flexible; it allows the incorporation of state-dependent detection and survival probabilities and uses the Bayesian framework [10]. However, there is no free lunch since the combinatorial explosion of the data association hypotheses that need to be considered for an optimal estimator still remains [13]. A suboptimal method, the probability hypothesis density (PHD) filer [9] is a widespread approximate implementation [14, 15].

To lessen the burden of the data association problem, the symmetric measurement equation (SME) approach is proposed by Kamen in [3, 4]. In short, it is intended to reduce the complexity of the data association by a factor of $N!$. Considering an MTT problem, the JPDAF can have $N!$ hypotheses for M measurements and $N = M$ targets, assuming no missed and false detections. The SME approach is to create symmetric functions that are invariant under the $N!$ permutations of the measurements and use these in the filtering process. The symmetric functions are usually highly nonlinear polynomials; thus, the SME approach turns the data association problem into a nonlinear estimation problem. The SME filter can reduce the complexity by $N!$ in the presence of false detections, too, because a variational problem has $M!/(M-N)!$ solutions while a combinatorial problem has $M!/(N!(M-N)!)$.

This paper investigates extended Kalman filter (EKF) based approaches to the SME multi-target tracking problem. The two common formalisms, the sum-of-products and the sum-of-powers are considered in tracking scenarios. It is found that the sum-of-powers EKF has a minimum in track loss regarding measurement noise levels. The general SME formalism is given in Sect. 2. The case study for three targets in the sum-of-products and sum-of-powers formalisms is discussed in Sect. 3. Numerical results from simulation experiments are presented in Sect. 4. Section 5 gives the conclusion and possible directions for further research.

2 Problem Statement

This section reviews the basics of the SME formulation of the multi-target state estimation task. It is assumed that all targets are detectable, no false measurements arrive, and the standard measurement model applies. Consider a tracking problem in one dimension where each target's state x evolves according to a discrete-time dynamic model of the form

$$x_{i,k+1} = f_k(x_{i,k}) + w_k \tag{2}$$

where f is the state transition function, w is an additive noise, and k is the discrete-time index. The function f and noise w may be different for each target. Let the measurement z be

$$z_{i,k} = x_{i,k} + \nu_k \tag{3}$$

where the additive term ν is a zero mean Gaussian random variable with variance σ_ν^2. Again, the noise terms may be different for each target. In case of N targets we collect N measurements at timestep k so we have a measurement vector $\mathbf{z} = \{z_1, z_2, \ldots z_N\}$. From now on the time index k will be omitted for brevity. To avoid the data association task of pairing the measurement with the predicted target tracks in the SME formulation, a new measurement vector

$$\mathbf{y} = h(\mathbf{z}) \tag{4}$$

is created that is invariant under the permutations of the z_i measurements in $\mathbf{z}$. The two main approaches to define the transformation function h are the sum-of-powers and sum-of-products formulations which create the following measurements, respectively:

$$\mathbf{y}_{pow} = \begin{bmatrix} \sum_i z_i \\ \sum_i z_i^2 \\ \vdots \\ \sum_i z_i^N \end{bmatrix}, \quad \mathbf{y}_{prod} = \begin{bmatrix} \sum_i z_i \\ \sum\limits_{1 \le i < j \le N} z_i z_j \\ \vdots \\ \prod_i z_i \end{bmatrix}. \tag{5}$$

Note that the elements of $\mathbf{y}_{prod}$ are the elementary symmetric polynomials in N variables up to degree N. All the elements in both $\mathbf{y}_{pow}$ and $\mathbf{y}_{prod}$ carry information about every target and are invariant to permutations; thus, no data association is needed. The downside is that the measurement model became nonlinear and non-Gaussian. Nonetheless, a Gaussian approximation is still possible in the form [4]:

$$\mathbf{y} = h(\mathbf{x}) + \mathbf{v}, \tag{6}$$

where $\mathbf{x}$ is the collection of the x_i state variables and $\mathbf{v}$ is a zero mean Gaussian random vector. The covariance matrix of $\mathbf{v}$ is constructed using the transformation function h and its derivatives.

3 Case Study for Three Objects

Consider a problem where three objects are tracked in a one-dimensional state. Let the states be x_1, x_2, and x_3. The original measurements are

$$z_{i,k} = x_{i,k} + \nu_k, \tag{7}$$

where the zero mean Gaussian noise term ν has standard deviation σ_ν. The sum-of-powers and sum-of-products measurement equations are

$$\mathbf{y}_{pow} = \begin{bmatrix} z_1 + z_2 + z_3 \\ z_1^2 + z_2^2 + z_3^2 - 3\sigma_\nu^2 \\ z_1^3 + z_2^3 + z_3^3 - 3\sigma_\nu^2(z_1 + z_2 + z_3) \end{bmatrix}, \quad \mathbf{y}_{prod} \begin{bmatrix} z_1 + z_2 + z_3 \\ z_1 z_2 + z_2 z_3 + z_1 z_3 \\ z_1 z_2 z_3 \end{bmatrix}. \tag{8}$$

To approximate the nonlinear and nongaussian equations (8) in the form (6) the noise vector $\mathbf{v}$ and its covariance matrix $\mathbf{R} = \mathrm{Cov}(\mathbf{v})$ should be determined. The subtracted terms in $\mathbf{y}_{pow}$ are to make the additive noise $\mathbf{v}$ a zero mean random vector. The derivation for the sum-or-products case is given in [5, Chap. 8.6] while results for the sum-of-powers case is published in [6]. For the sum-of-product formulation, the covariance of the transformed noise vector is

$$\mathbf{R}_{prod} = 3\sigma_\nu^2 \mathbf{V}_1 \mathbf{V}_1^\top + 3\sigma_\nu^4 \mathbf{V}_2 \mathbf{V}_2^\top + \sigma_\nu^6 \mathbf{V}_3 \mathbf{V}_3^\top, \tag{9}$$

where

$$\mathbf{V}_1 = \begin{pmatrix} 1 & 1 & 1 \\ x_2 + x_3 & x_1 + x_3 & x_1 + x_2 \\ x_2 x_3 & x_1 x_3 & x_1 x_2 \end{pmatrix}, \mathbf{V}_2 = \begin{pmatrix} 0 & 0 & 0 \\ 1 & 1 & 1 \\ x_3 & x_2 & x_1 \end{pmatrix}, \mathbf{V}_3 = \begin{pmatrix} 0 \\ 0 \\ 1 \end{pmatrix}. \tag{10}$$

For the sum-of-powers case the covariance matrix is

$$\mathbf{R}_{pow} = \begin{bmatrix} 3\sigma_\nu^2 & 2\sigma_\nu^2 s_1 & 3\sigma_\nu^2 s_2 + 9\sigma_\nu^4 \\ 2\sigma_\nu^2 s_1 & 4\sigma_\nu^2 s_2 + 6\sigma_\nu^4 & 6\sigma_\nu^2 s_3 + 12\sigma_\nu^4 s_1 \\ 3\sigma_\nu^2 s_2 + 9\sigma_\nu^4 & 6\sigma_\nu^2 s_3 + 12\sigma_\nu^4 s_1 & 9\sigma_\nu^2 s_4 + 36\sigma_\nu^4 s_2 + 45\sigma_\nu^6 \end{bmatrix}, \tag{11}$$

where $s_i = x_1^i + x_2^i + x_3^i$.

The Jacobian matrices of the transformation functions h_{prod} and h_{pow} are

$$\mathbf{H}_{prod} = \begin{bmatrix} 1 & 1 & 1 \\ x_2 + x_3 & x_1 + x_3 & x_1 + x_2 \\ x_2 x_3 & x_1 x_3 & x_1 x_2 \end{bmatrix}, \mathbf{H}_{pow} = \begin{bmatrix} 1 & 1 & 1 \\ 2x_1 & 2x_3 & 2x3 \\ 3x_1^2 - 3\sigma_\nu^2 & 3x_2^2 - 3\sigma_\nu^2 & 3x_3^2 - 3\sigma_\nu^2 \end{bmatrix}. \tag{12}$$

Using the above ingredients, an extended Kalman filter recursion can be constructed for the tracking problem using symmetric measurement equations in either of the two discussed formalisms. Let the vector $\hat{\mathbf{x}}_{k|k-1}$ denote the prediction of the three-object coordinates, and $\mathbf{P}_{k|k-1}$ the predicted error covariance matrix. For the measurement update step, the following applies:

$$\mathbf{P}_{k|k} = \mathbf{P}_{k|k-1} - \mathbf{K}_k \mathbf{H}_k \mathbf{P}_{k|k} \tag{13}$$

$$\hat{\mathbf{x}}_{k|k} = \hat{\mathbf{x}}_{k|k-1} + \mathbf{K}_k (y_k - h(\hat{\mathbf{x}}_{k|k-1})) \tag{14}$$

$$\mathbf{K}_k = \mathbf{P}_{k|k-1} \mathbf{H}_k^\top (\mathbf{H}_k \mathbf{P}_{k|k-1} \mathbf{H}_k^\top + \mathbf{R}_k)^{-1} \tag{15}$$

where $\mathbf{h}$ is from (8), $\mathbf{H}$ is from (12), and $\mathbf{R}$ is from (9) or (11). The predicted state $\hat{\mathbf{x}}_{k|k-1}$ is substituted into $\mathbf{H}$ and $\mathbf{R}$.

4 Example

A numerical experiment is conducted for a three-target setup to evaluate the sum-of-products and sum-of-powers EKFs. The discrete-time system model is

$$x_{i,k+1} = x_{i,k} + \alpha_i T + w_k \tag{16}$$

$$z_{i,k} = x_{i,k} + \nu_k\,, \tag{17}$$

where α_i is a speed value, T is the sampling time and w is a zero mean Gaussian noise.

In the simulation, the initial position of the targets is $\mathbf{x}_0 = [80, -10, 10]^\top$. The speed values are $\alpha = [-1, +0.6, -0.2]$ and $T = 1$. The process noise variance is $q = 1$, and the measurement noise variance is $\sigma_v^2 = 100$; A realization of trajectories over a 120 timestep period is shown in Fig. 2. The large noise on the measurements and the close spacing of the objects make this scenario a difficult problem for a multi-target tracker. Besides the SME EKF filters, a matched KF is used. This filter always gets the appropriate measurement for a given track; thus, it can provide a theoretical minimum for the estimation error covariance.

200 Monte Carlo runs were performed for the three filter configurations. Figure 3 shows the results for the matched KF. The trajectory plot presents one realization of the problem. The error plot shows 10% of the 200 error curves along with the mean and one sigma band. The matched Kalman filter gives a balanced performance.

The SME-based EKF solutions suffer occasional track loss. An example of the sum-of-product approach is shown in Fig. 4, where the tracks of x_1 and x_3 are swapped just before the actual crossing around timestep 90. In this case, the estimation recovers because of the real track crossing; however, this does not always happen. Indeed, from the 200 trials, the estimation diverges 66 times for the sum-of-product and 65 times for the sum-of-powers filters. For the creation of the error plots, these estimations are omitted in Figs. 4 and 5, and the root mean squared error values are calculated accordingly.

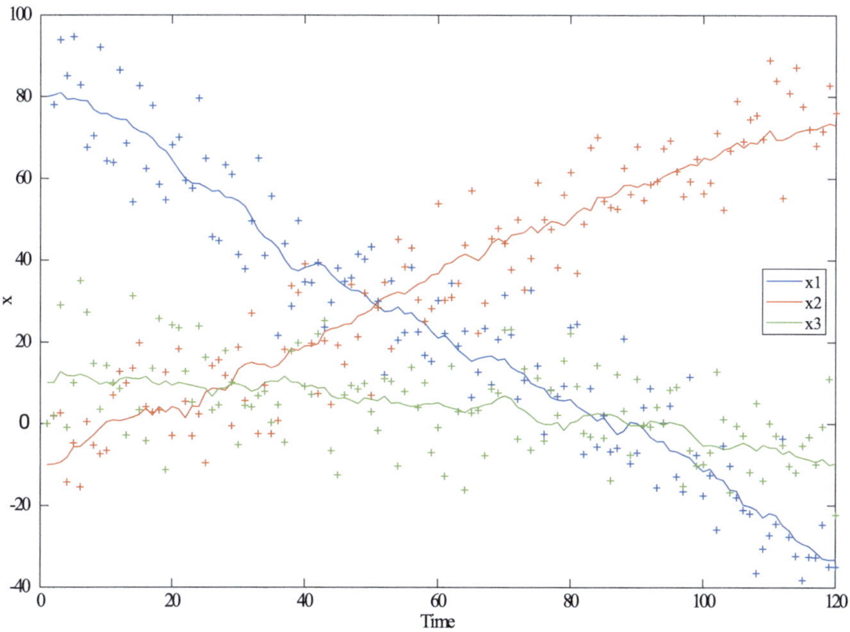

Fig. 2 Ground truth trajectories and measurements for a scenario

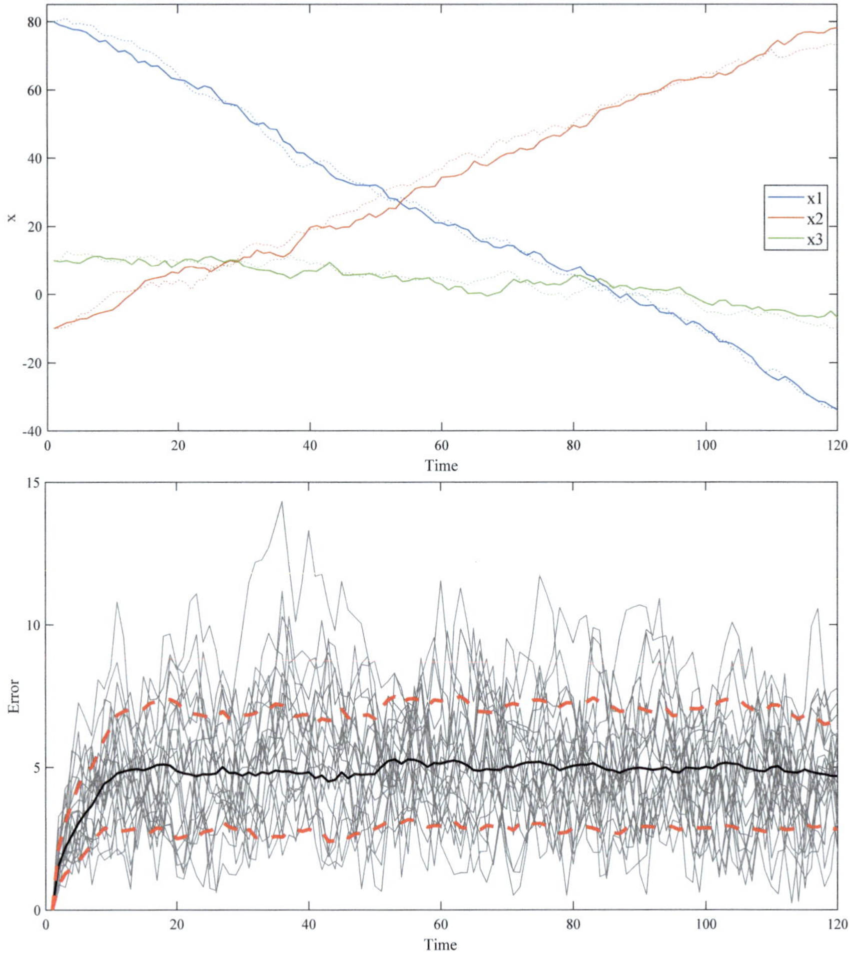

Fig. 3 Results for the matched Kalman filter

Simulations were run with different measurement noise intensity levels. The results are summarized in Table 1. Interestingly, the sum-of-powers approach has a minimum at noise level 10 in terms of track loss. At the high noise level end, the two approaches converge to a common value. This feels natural since the measurements are less informative at this noise level. The rmse values for the sum-of-powers approach have lower values which indicate that the sum-of-products approach is more unstable in this scenario. What is particular in the current setup is that the noise has a strongly nonlinear effect on the sum-of-powers EKF formulation.

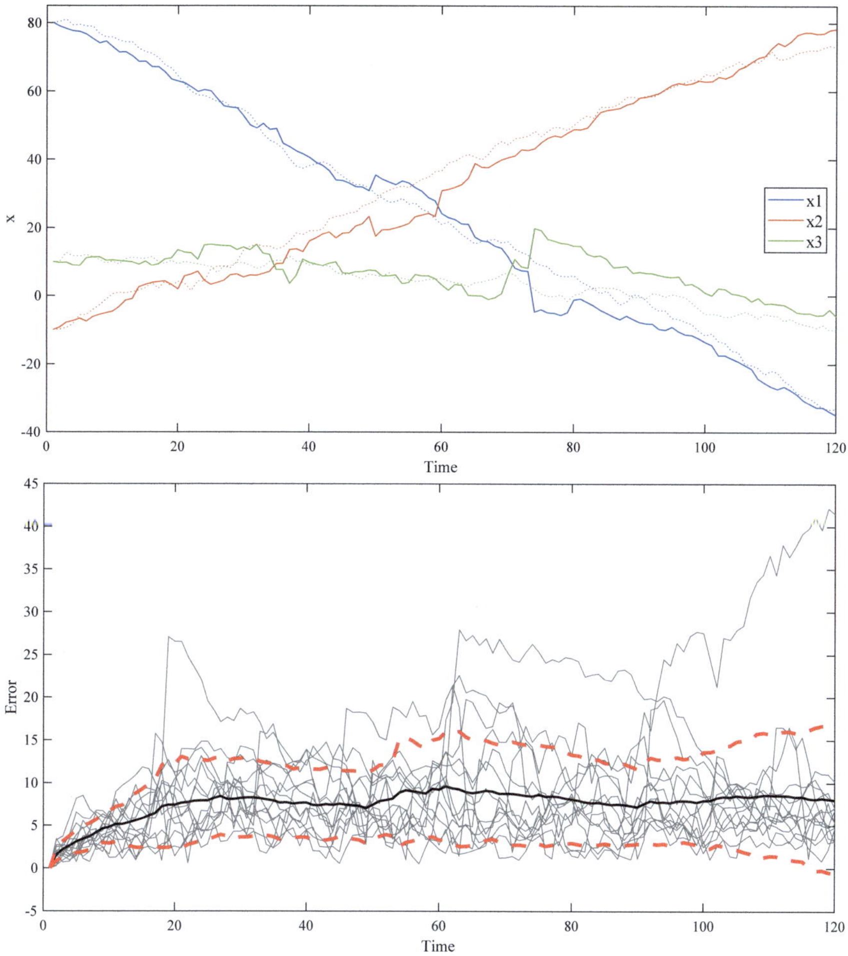

Fig. 4 Results for the sum-of-products EKF

5 Conclusion

The sum-of-product and sum-of-powers symmetric measurement equations are applied to the multi-target tracking problem. In a three-object scenario, the performances of these approaches are compared using various measurement noise levels. It is found that the sum-of-powers formalism has a minimum in terms of track loss around the process-to-measurement noise level ratio of 1/10. Further investigations are needed to assess the impact of noise level ratios and system dynamics on the filtering performances of the different approaches.

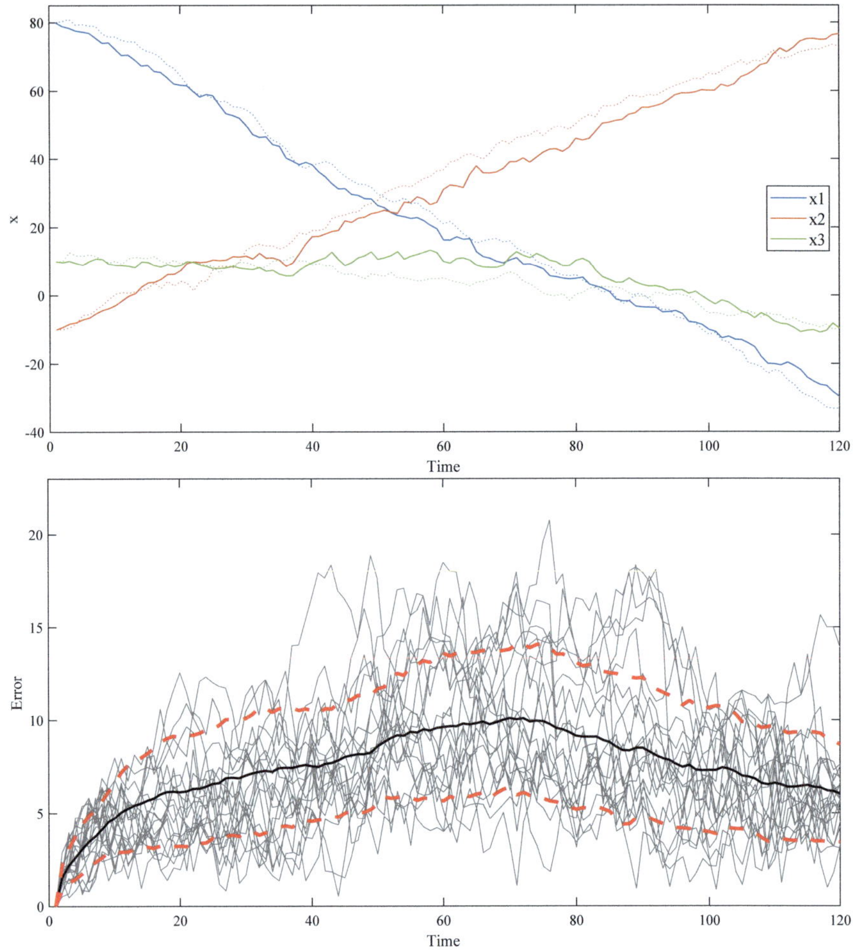

Fig. 5 Results for the sum-of-powers EKF

Table 1 Results of 200 MC trials for different measurement noise intensity levels

Measurement noise intensity (σ_v^2)	1	5	10	50	100
Matched KF rmse	1.36	2.32	2.84	4.39	5.23
Sum-of-products EKF rmse	4.32	4.43	5.26	7.04	9.34
Sum-of-powers EKF rmse	2.54	2.80	3.45	6.17	8.33
Sum-of-products EKF track loss	44	54	53	67	66
Sum-of-powers EKF track loss	74	18	8	65	65

Acknowledgements This research was supported by the European Union within the framework of the National Laboratory for Autonomous Systems (RRF-2.3.1-21-2022-00002).

References

1. Bar-Shalom Y, Daum F, Huang J (2009) The probabilistic data association filter. IEEE Control Syst Mag 29(6):82–100
2. Fortmann TE, Bar-Shalom Y, Scheffe M (1980) Multi-target tracking using joint probabilistic data association. In: 1980 19th ieee conference on decision and control including the symposium on adaptive processes. IEEE, pp 807–812
3. Kamen E (1989) Multiple target tracking based on symmetric measurement equations. In: 1989 American control conference. IEEE, pp 2690–2695
4. Kamen E (1992) Multiple target tracking based on symmetric measurement equations. IEEE Trans Autom Control 37(3):371–374
5. Kamen EW, Su JK (1999) Introduction to optimal estimation. Springer Science & Business Media
6. Leven WF, Lanterman AD (2009) Unscented kalman filters for multiple target tracking with symmetric measurement equations. IEEE Trans Autom Control 54(2):370–375
7. Lindenmaier L, Aradi S, Bécsi T, Törő O, Gáspár P (2022) GM-PHD filter based sensor data fusion for automotive frontal perception system. IEEE Trans Veh Technol 71(7):7215–7229
8. Lindenmaier L, Aradi S, Bécsi T, Törő O, Gáspár P (2023) Object-level data-driven sensor simulation for automotive environment perception. IEEE Trans Intell Vehicles
9. Mahler R (2003) Multitarget bayes filtering via first-order multitarget moments. IEEE Trans Aerosp Electron Syst 39(4):1152–1178
10. Mahler R (2013) "Statistics 102" for multisource-multitarget detection and tracking. IEEE J Sel Topics Signal Process 7(3):376–389
11. Mahler RP (2014) Advances in statistical multisource-multitarget information fusion. Artech House
12. Richards MA (2014) Fundamentals of radar signal processing. McGraw-Hill Education
13. Ristic B, Beard M, Fantacci C (2016) An overview of particle methods for random finite set models. Inf Fusion 31:110–126
14. Vo BN (2005) Sequential Monte Carlo methods for multitarget filtering with random finite sets. IEEE Trans Aerosp Electron Syst 41(4):1224–1245
15. Vo BN, Ma WK (2006) The Gaussian mixture probability hypothesis density filter. IEEE Trans Signal Process 54(11):4091–4104
16. Vo Bn, Mallick M, Bar-Shalom Y, Coraluppi S, Osborne R, Mahler R, Vo Bt (2015) Multitarget tracking. Wiley encyclopedia of electrical and electronics engineering

Multi-agent Reinforcement Learning-Based Traffic Control for Urban Intersections Using Variable Speed Limit Control

Xuan Fang and Tamás Tettamanti

1 Introduction

The intersection is the basic unit, core, and hub of the urban road network. However, intersections are also bottlenecks in road capacity and frequent locations for traffic congestion and traffic accidents. The rapid growth in the number of vehicles far exceeds the capacity of existing road infrastructure, which leads to increased delays and frequent accidents at intersections. Research shows that most urban road accidents occur at intersections [1, 2]. Therefore, the management and optimization of intersections are key to managing the urban transportation system.

The commonly used urban traffic management strategy is to optimize the signal control of intersections, which can achieve higher traffic throughput to reduce congestion and improve the capacity of high-load urban networks. With the revolutionary development of computers, communications, and traffic detection technologies, the technical environment of urban traffic signal control systems is evolving from experience-driven to data-driven control. Multi-agent reinforcement learning (MARL) has developed rapidly in urban traffic signal control and has been applied in practical applications. Marco Wiering used the multi-agent reinforcement learning algorithm for learning traffic light controllers to minimize the overall waiting time of travelers [3]. Prabuchandran K.J. et al. utilized multi-agent Q-learning on each traffic signal junction to minimize the average delay of the vehicles in the network [4]. Rather than the mature Q-learning algorithm, Tong Wu et al. proposed a multi-agent recurrent deep deterministic policy gradient (MARDDPG) algorithm for traffic light control to minimize not only vehicles but also pedestrians waiting to pass through the intersection [5]. Tianshu Chu et al. applied an advantage actor-critic(A2C) based

X. Fang (✉) · T. Tettamanti
Department of Control for Transportation and Vehicle Systems, Faculty of Transportation Engineering and Vehicle Engineering, Budapest University of Technology and Economics, Budapest, Hungary
e-mail: fangxuan@edu.bme.hu

G. L. Conte and O. Sename (eds.), *Proceedings of 12th International Conference on Mechatronics and Control Engineering*, Lecture Notes in Mechanical Engineering,
https://doi.org/10.1007/978-981-96-6452-8_11

MARL framework on a large synthetic traffic grid and a large real-world traffic network of Monaco city. Simulation results demonstrated the effectiveness of the proposed control method in reducing average queue length and delay [6]. However, considering the massive financial budget required to upgrade the huge number of existing conventional traffic lights to adaptive ones, most traffic lights are still controlled with the predefined fixed-time scheme [7]. Moreover, although traffic signal control can help improve intersection capacity, it still cannot eliminate complete stops and oscillating of vehicles, which pose safety hazards. There are intersections where the constraint of the tram schedule is primarily, i.e., traffic lights are coordinated but fixed at all times in order to serve the tram schedule. Applying RL-based traffic signal control in such an intersection is not feasible.

The emergence of Connected and Autonomous Vehicles (CAV) provides a new perspective for urban intersection control, that is, the perspective of the vehicle rather than the traffic infrastructure. As an intelligent traffic management system, the Variable speed limit (VSL) control strategy has been widely applied on freeways to improve traffic efficiency [8, 9] and prevent traffic accidents [10]. VSL control can also be used to reduce environmental impacts by reducing stop-and-go phenomena [11].

The primary objective of this study was to design a MARL-based VSL control framework at the urban intersection to smooth traffic and improve traffic safety. The proposed VSL control framework is tested on the microscopic traffic simulator simulation of urban mobility (SUMO). The traffic safety assessment is completed using the Surrogate Safety Assessment Model (SSAM).

2 Methodology

This section introduces the proposed MARL-based VSL control framework. Section 2.1 introduces the background of Reinforcement Learning (RL) and the multi-agent proximal policy optimization (MAPPO) algorithm utilized in this study. Section 2.2 introduces the Surrogate Safety Assessment Model (SSAM) used to analyze traffic conflicts.

2.1 Multi-agent Reinforcement Learning

Reinforcement learning is a learning mechanism that performs decision-making through the interaction between the Agent and the dynamic environment. It is a trial-and-error learning method. Trial and delayed rewards are two of the most important properties of reinforcement learning. The basic principle of reinforcement learning is the Agent performs a certain action on the environment, changes the state of the environment, and obtains the reward signal given by the environment to strengthen the mapping relationship between a certain state and the best action strategy. By repeatedly executing this process, the Agent can acquire the ability to give the best action strategy in any environmental state.

The proximal policy optimization (PPO) algorithm is an improved method based on the policy gradient (PG) algorithm [12] and the trust region policy optimization (TRPO) algorithm [13]. The PPO algorithm solves the weakness of the PG algorithm's poor robustness and the TRPO algorithm's complexity. In the PPO algorithm, the Agent is also called Actor. Actor has its own policy π for a specific task. Strategy π is usually represented by a neural network with parameters θ. The objective function of the Actor is defined as

$$\max_{\theta} J^{clip}(\theta) = \widehat{E}_t \left[clip\left(\frac{\pi_\theta(a_t|s_t)}{\pi_{\theta_{old}}(a_t|s_t)}, 1-\varepsilon, 1+\varepsilon \right) \widehat{A}_t, \frac{\pi_\theta(a_t|s_t)}{\pi_{\theta_{old}}(a_t|s_t)} \widehat{A}_t \right] \quad (1)$$

where $\widehat{E}_t$ is the unbiased estimate of the gradient. π_θ and $\pi_{\theta_{old}}$ are new and old strategies network used in the TRPO algorithm, which allows the network to interact with the environment and use the collected data to train π_θ. This helps the agent interact with the environment once and perform multiple parameter updates, increasing the update speed. The Clip function is used to limit the probability ratio $R_t(\theta)$ to $(1-\varepsilon, 1+\varepsilon)$, where $R_t(\theta) = \frac{\pi_\theta(a_t|s_t)}{\pi_{\theta_{old}}(a_t|s_t)}$, ε is a hyperparameter, also take $\frac{\pi_\theta(a_t|s_t)}{\pi_{\theta_{old}}(a_t|s_t)} \widehat{A}_t$ as the optimization object. when $\widehat{A}_t > 0$, increase the probability of $\pi_\theta(a_t|s_t)$, vice versa. a_t is action space and s_t is state space. $\widehat{A}_t$ is the advantage function estimated by critic. The optimization framework of the PPO algorithm for urban intersection VSL control is shown in Fig. 1.

This study uses the mainstream Centralized Training with Decentralized Execution (CTDE) mechanism to train a centralized Critic network with global information to estimate the state value and advantage function of each agent, extending the PPO algorithm to the MARL.

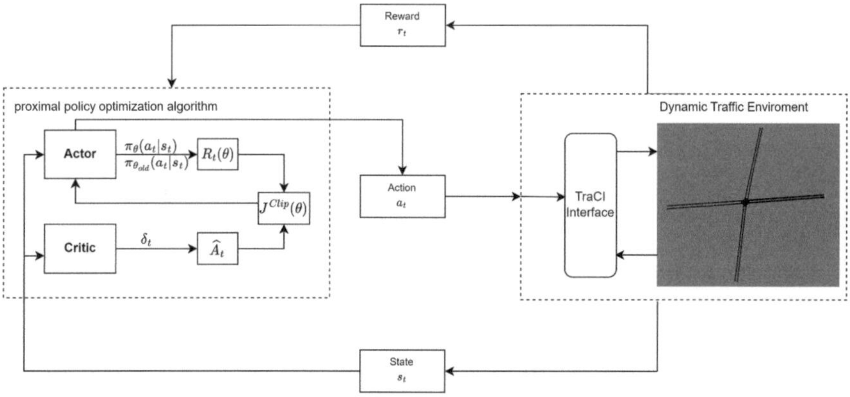

Fig. 1 Reinforcement learning-based intersection control framework

2.2 Surrogate Safety Assessment Model

Road safety evaluation methods are divided into direct evaluation methods and indirect evaluation methods. The direct evaluation method based on traffic accident statistics is widely used. The shortage is insufficient traffic accident data. Based on traffic conflict technology, Douglas Gettman and Larry Head [14] investigated the possibility of deriving alternative safety measures from existing microscopic traffic simulation models, known as the Surrogate Safety Assessment Model (SSAM).

The main surrogate safety measures of SSAM are the time to collision (TTC), the post-encroachment time (PET), maximum speed (MaxS), the speed differential (DeltaS), and maximum acceleration (MaxD). PET refers to the time difference between the leading vehicle passing a certain position and the following vehicle passing the same position. MaxS refers to the maximum speed of the two vehicles during the entire conflict. DeltaS refers to the relative speed of the two vehicles when the driver takes avoidance measures. MaxD refers to the maximum acceleration or deceleration speed of the following vehicle during the conflict. In the SSAM model, traffic conflicts can be divided into rear-end collisions, lane changes, and crossing conflicts according to the relative angles of conflicting vehicles.

SSAM combines the vehicle trajectory files exported from microscopic traffic simulation (SUMO, VISSIM, AIMSUM, PARAMICS, etc.) with the threshold of the conflict surrogate safety measures to determine the traffic conflicting status. When the minimum TTC and PET values exceed predetermined thresholds, a conflict is recorded in SSAM, and the type of conflict is determined based on the lane and the angle between the two adjacent vehicles [15]. Research shows that SSAM is an effective method to evaluate traffic safety [16, 17].

3 Case Study

The dynamic traffic environment is created in the open-source microscopic traffic simulator SUMO. The traffic state, reward value, and the agent's action strategy are transmitted through the embedded traffic control interface (TraCI). Taking the József körút (József boulevard) - Üllöi út (Üllöi Road) intersection in Budapest as the experimental scenario of this study. Üllöi Road is the longest urban road in Hungary. As part of the Outer Ring Road, József Boulevard is one of the busiest roads in Budapest. The geometric structure of the simulated intersection is obtained through Open Street Map (OSM) and converted into a road network in SUMO. Figure 2 shows the geometric structure of the simulated intersection.

As the impact of the introduction of CAVs on travel demand is uncertain, a set of random trips with a 0.8 s repetition rate is generated. Road sections in the four directions are set to be 500 m long, respectively. The simulation time is set to 7200 s. The fixed cycle traffic light control schema exists at the intersection. The cycle time

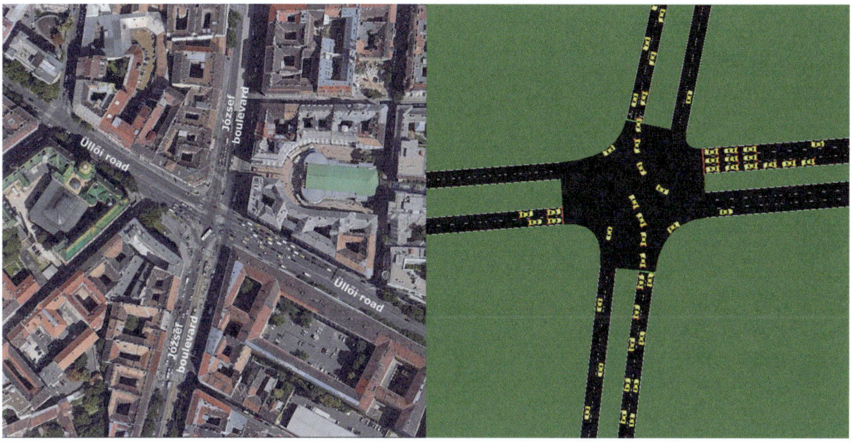

Fig. 2 Geometric structure of the simulated intersection. The GPS coordinates are 47.4858° N, 19.0700° E

is 90 s with 60 s of green phase, the total time allocated for all conflicting directions that have the green signal. The 3-s yellow phase serves as a transition between the green and red phases.

The critical elements in the MARL include the state design, action set, and reward function. State definition of traffic control problems has no clear agreement on the form. It highly depends on traffic data the vehicle and infrastructure collected. The commonly used state representations in RL-based intersection control are vehicle numbers [18, 19], queue length [20–22], and cumulative delay [23, 24]. The environment state in this paper is the queue length in each entrance section. The discrete speed limit was designed as the agent's action in the MARL-based VSL control. The set of actions is [16.67, 13.89, 11.11, 8.3, 5.55], and the unit is m/s. The goal of individual agents is to cooperate with each other to find the optimal speed limits set. Reward is the only feedback from the environment after the agent performs actions. The objective function of the MARL-based VSL control is the negative of the standard deviation of the vehicle occupancy data on different road sections.

The choice of hyperparameters directly affects the performance of the algorithm model. There are no default hyperparameters in reinforcement learning, but a combination of hyperparameters needs to be set for different tasks. Table 1 shows the set of MAPPO hyperparameters utilized in this paper.

4 Results and Discussion

The Min-max normalized reward curve verifies the training performance through 600 iterations. In each iteration, a random seed is applied for the proposed control strategies to keep the diversity. Figure 3 shows the negative standard deviation of the

Table 1 Multi-agent proximal policy optimization hyperparameters for training

Hyperparameters	Value
Number of training iterations	600
Learning rate	0.0005
Number of agents	4
Discount factor γ	0.99
GAE parameter λ	0.95
Time step per update	120
Number of PPO epochs per update	15
Structure of neural network	$64 \times 64 \times 64$
Hidden layer activation function	ReLU

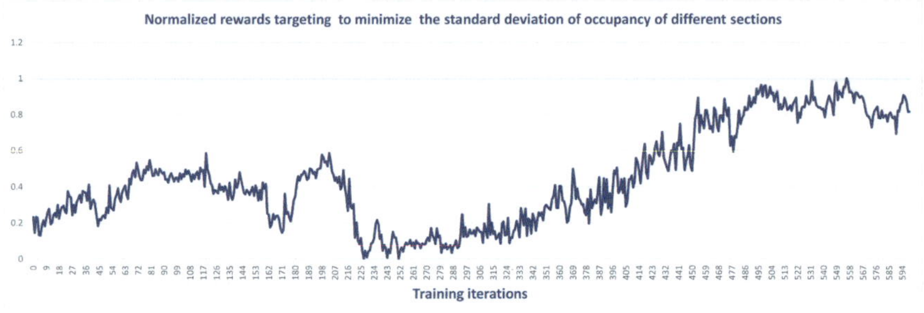

Fig. 3 Convergence of the MAPPO algorithm

occupancy of different road sections. It can be seen from the figure that in the early stage of training because the agent has been exploring the optimal control strategy, the reward fluctuates. After the agent trained for 250 episodes, the algorithm began to converge, indicating that the agent gradually became familiar with the simulation environment and could perform the optimal speed limit set based on the reward.

Figure 4 shows the spatiotemporal distribution of road section occupancy to illustrate the control effect intuitively. It can be seen intuitively from the figure that the MARL-based VSL control has greatly reduced the number of vehicles queuing on the road. Except for the southbound lane, which still has a high lane occupancy, the occupancy on the roads in the other three directions remained at a low level, and vehicles are distributed uniformly on them. The relatively low road occupancy alleviates the stop-and-go phenomenon and the possibility of traffic conflicts. The high occupancy of the southbound road can be explained by the limitations of the road infrastructure. About 200 m past the intersection, the number of southbound lanes merges from two to a single traffic lane, resulting in continued congestion of southbound traffic.

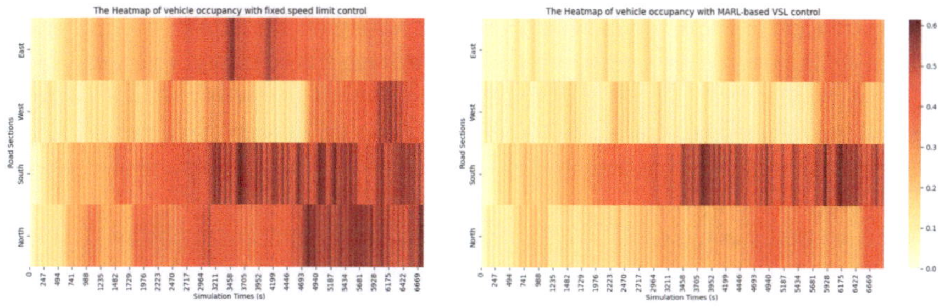

Fig. 4 The spatiotemporal distribution of road section occupancy

Table 2 SSAM simulation experiment results

Control scheme	Total number of traffic conflicts	Crossing	Rear end	Lane change	MaxD (m/s)
Fixed speed limit	5952	541	2856	2555	3.36–20.6
MARL-based VSL	5393	583	2599	2211	3.26–12.82
Relative value	−9.39%	+7.76%	−8.99%	−13.46%	−3%–60.69%

According to the current traffic safety strategy applied in the study area, the maximum speed limit was reduced from 50 km/h to 40 km/h. Based on the control output, the trajectory file of the simulated vehicle is imported into the SSAM software. Table 2 shows the comparison of the conflicts applying the fixed speed limit control and MARL-based VSL control. The SSAM results demonstrated that compared to the current traffic safety strategy, although the risk of crossing has increased, the MARL-based VSL control reduces total traffic conflicts by 9.39% at the designated intersections. Furthermore, the more minor acceleration variation of the following vehicle during the conflict confirms the effectiveness of the proposed control framework in smoothing traffic.

5 Conclusion

This article proposes a new perspective on multi-agent reinforcement learning-based urban intersection control. Instead of centralized traffic signal control, this paper implements distributed variable speed limit control utilizing CAVs. The proposed control framework implements four individual agents on four entrances of the intersection. These four individual agents work fully cooperatively to smooth traffic and reduce traffic conflicts. A real-world urban intersection is generated in the microscopic traffic simulator SUMO based on the OSM data. The spatiotemporal distribution of road section occupancy intuitively shows improvements in the overall vehicle distribution and traffic conflicts. The SSAM simulation results demonstrate the effectiveness of the proposed control method in improving traffic safety, with total traffic conflicts reduced by 9.39% at the designated intersection. Previous research work has verified the effectiveness of MARL-based VSL control in reducing greenhouse gas emissions and total waiting time in highway scenarios. Future research will combine the highway and the urban scenario VSL control to achieve regional traffic control to alleviate traffic congestion and improve traffic safety and sustainability.

Acknowledgment The research was supported by the European Union within the framework of the National Laboratory for Autonomous Systems (RRF-2.3.1- 21-2022-00002).

References

1. Archer J, Vogel K (2000) The traffic safety problem in urban areas. Res, Rep
2. Haleem K, Abdel-Aty M, Mackie K (2010) Using a reliability process to reduce uncertainty in predicting crashes at unsignalized intersections. Accid Anal Prevent 42(2):654–666
3. Wiering MA et al (2000) Multi-agent reinforcement learning for traffic light control. In: Machine learning: proceedings of the seventeenth international conference (ICML'2000), pp 1151–1158
4. Prabuchandran K, AN HK, Bhatnagar S (2014) Multi-agent reinforcement learning for traffic signal control. In: 17th international IEEE conference on intelligent transportation systems (ITSC). IEEE, pp 2529–2534
5. Wu T, Zhou P, Liu K, Yuan Y, Wang X, Huang H, Wu DO (2020) Multi-agent deep reinforcement learning for urban traffic light control in vehicular networks. IEEE Trans Veh Technol 69(8):8243–8256
6. Chu T, Wang J, Codecà L, Li Z (2019) Multi-agent deep reinforcement learning for large-scale traffic signal control. IEEE Trans Intell Transp Syst 21(3):1086–1095
7. Lämmer S, Helbing D (2008) Self-control of traffic lights and vehicle flows in urban road networks. J Stat Mech: Theory Exp 2008(04):P04019
8. Han Y, Hegyi A, Zhang L, He Z, Chung E, Liu P (2022) A new reinforcement learning-based variable speed limit control approach to improve traffic efficiency against freeway jam waves. Transp Res Part C: Emerg Technol 144:103900
9. Wu Y, Tan H, Qin L, Ran B (2020) Differential variable speed limits control for freeway recurrent bottlenecks via deep actor-critic algorithm. Transp Res Part C: Emerg Technol 117:102649
10. Li Z, Xu C, Guo Y, Liu P, Pu Z (2020) Reinforcement learning-based variable speed limits control to reduce crash risks near traffic oscillations on freeways. IEEE Intell Transp Syst Mag 13(4):64–70
11. Zhu F, Ukkusuri SV (2018) Modeling the proactive driving behavior of connected vehicles: a cell-based simulation approach. Comput-Aided Civil Infrastruct Eng 33(4):262–281
12. Sutton RS, McAllester D, Singh S, Mansour Y (1999) Policy gradient methods for reinforcement learning with function approximation. In: Advances in neural information processing systems, vol 12
13. Schulman J, Levine S, Abbeel P, Jordan M, Moritz P (2015) Trust region policy optimization. In: International conference on machine learning. PMLR, pp 1889–1897
14. Gettman D, Head L (2003) Surrogate safety measures from traffic simulation models. Transp Res Rec 1840(1):104–115
15. Zhang L (2017) Open source SSAM user manual. https://github.com/OSADP/SSAM/blob/master/Documentation/Open%20Source%20SSAM%20User%20Manual.pdf

16. Huang F, Liu P, Yu H, Wang W (2013) Identifying if vissim simulation model and ssam provide reasonable estimates for field measured traffic conflicts at signalized intersections. Accid Anal Prevent 50:1014–1024
17. Wu J, Radwan E, Abou-Senna H (2018) Determination if vissim and ssam could estimate pedestrian-vehicle conflicts at signalized intersections. J Transp Saf Sec 10(6):572–585
18. Thorpe TL, Anderson CW (1996) Tra c light control using sarsa with three state representations. Technical report, Citeseer
19. Wen K, Qu S, Zhang Y (2007) A stochastic adaptive control model for isolated intersections. In: 2007 IEEE international conference on robotics and biomimetics (ROBIO). IEEE, pp 2256–2260
20. Abdulhai B, Pringle R, Karakoulas GJ (2003) Reinforcement learning for true adaptive traffic signal control. J Transp Eng 129(3):278–285
21. El-Tantawy S, Abdulhai B (2020) An agent-based learning towards decentralized and coordinated traffic signal control. In: 13th International IEEE conference on intelligent transportation systems. IEEE, pp 665–670
22. El-Tantawy S, Abdulhai B, Abdelgawad H (2014) Design of reinforcement learning parameters for seamless application of adaptive traffic signal control. J Intell Transp Syst 18(3):227–245
23. Shoufeng L, Ximin L, Shiqiang D (2008) Q-learning for adaptive traffic signal control based on delay minimization strategy. In: 2008 IEEE international conference on networking, sensing and control. IEEE, pp 687–691
24. Touhbi S, Babram MA, Nguyen-Huu T, Marilleau N, Hbid ML, Cambier C, Stinckwich S (2017) Adaptive traffic signal control: exploring reward definition for reinforcement learning. Proc Comput Sci 109:513–520

Identification of Lateral Dynamics of an Autonomous Car Vehicle

Dániel Pup, György Istenes, Ferenc Szauter, and József Bokor

1 Introduction

Vehicle dynamics, modeling, and vibration analysis play an important role in the design of vehicle dynamics, ride comfort, and control. The design of driver assistance systems and autonomous vehicle functions requires the use of model-based control systems. This requires both the use of models developed by vehicle mechanics and their verification by measurements, and modeling from the measurements themselves, i.e., the application of system ID.

The measurements used for ID in this paper were made using a measurement and control system developed at the Széchenyi István University.

The starting points for modeling are the information about the real physical system can be theoretical knowledge (first principle modeling) or practical knowledge and assumptions.

The dynamic model describes how the characteristics of the system under study change over time which is usually an ordinary or partial differential equation, or system of equations. It is possible to describe, e.g., some conveniently chosen transformed domain (frequency domain).

In this paper, the lateral dynamics of a road vehicle are modeled using physical considerations, both for simulation and control purposes.

There are many studies on lateral vehicle dynamics, ranging from the simplest kinematic and bicycle-type models to high-dimensional multi-body descriptions, (see [1–3]). For testing control-oriented modeling methods, the choice of a medium complexity DT (Double Track) nonlinear model is the most appropriate: the corresponding lateral vehicle dynamics model has two translational and three rotational

D. Pup · G. Istenes (✉) · F. Szauter · J. Bokor
Széchenyi István University, Györ, Hungary
e-mail: istenes.gyorgy@ga.sze.hu

J. Bokor
Institute for Computer Science and Control (SZTAKI), Budapest, Hungary

G. L. Conte and O. Sename (eds.), *Proceedings of 12th International Conference on Mechatronics and Control Engineering*, Lecture Notes in Mechanical Engineering,
https://doi.org/10.1007/978-981-96-6452-8_12

degrees of freedom. The suspension system is modeled as a spring-damping system, where the rolling and pitching motions at the wheels are responsible for the load transfer. The intended use of the model-based control is to allow normal driving conditions and low wheel slip, such that the nonlinear wheel dynamics and tire grip characteristics are negligible in the model. This paper will therefore focus on the ID of linear models.

2 Test Vehicle and the Experimental Conditions

The Nissan Leaf is an electric-powered hatchback. Developed and launched by Nissan in 2010. The Nissan Leaf is the first fully electric car to be sold in large numbers. The test vehicle can be seen in Fig. 1.

The modification of the test vehicle and its sensor system was a team effort, carried out by colleagues from the Vehicle Industry Research Center of Széchenyi István University and SZTAKI. During the modification of the vehicle, a safety system, upper and lower level control system, measurement data acquisition system, and sensor system were installed. The sampling frequency was 0.02 Hz in every experiments.

The complete vehicle control system consists of a lower and an upper level control system. The high-level controller implements the trajectory determined by the autonomous function, while the low-level controller controls the steering servo. The block diagram of the vehicle control system can be seen in Fig. 2, and the description of the signs in Fig. 2 is shown in Table 1.

Fig. 1 The test vehicle

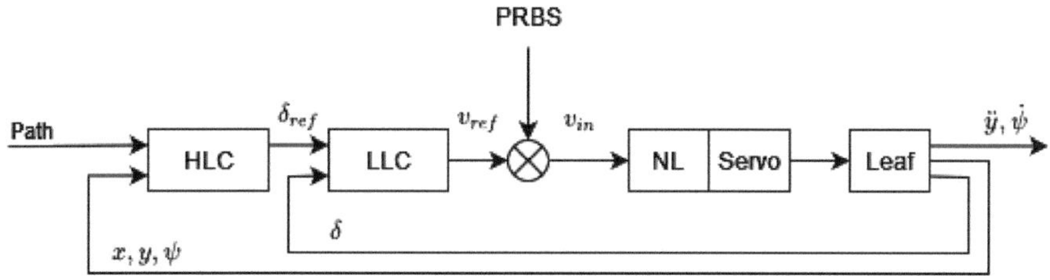

Fig. 2 The block diagram of the vehicle control system

Table 1 Description of the signs in Fig. 2

Sign	Description
HLC	High-level control: Controls the autonomous functions
LLC	Low-level control: Controls the required steering angle
NL	The nonlinearity of the steering servo
Servo	Steering servo
VD	Vehicle dynamic
Path	The vehicle's route
PRBS	This is a pseudo-random binary sequence
δ_{ref}	The steering angle reference
v_{ref}	The voltage reference
v_{in}	Voltage
δ	Steering angle
x, y	The vehicle position
ψ	The vehicle heading
$\ddot{y}$	Lateral acceleration
$\dot{\psi}$	Yaw rate

The measurements required to identify the lateral dynamics of the test vehicle were carried out at the ZalaZone test complex. The car followed pre-designed straight tracks at various but fixed speeds. The existing two-level control system ensures path and speed tracking in fully autonomous mode. The appropriate excitation for the modeling is provided by an artificial excitation signal added to the input signal of the low-level control.

For ID, the system is excited with a filtered PRBS (Pseudo Random Binary Sequence). The APSD (Auto Power Spectral Density) of PRBS is constant within a given frequency range and provides a signal with maximum energy under amplitude constraints. This external excitation signal is needed because without this, the vehicle dynamics wouldn't be properly excited, and thus ID of the system wouldn't be possible. The PRBS in time and frequency domain can be seen in Fig. 3.

According to the detailed model constructed from first principles physical, there is a resonance frequency of around 3.2 Hz. To accurately identify this, it would be

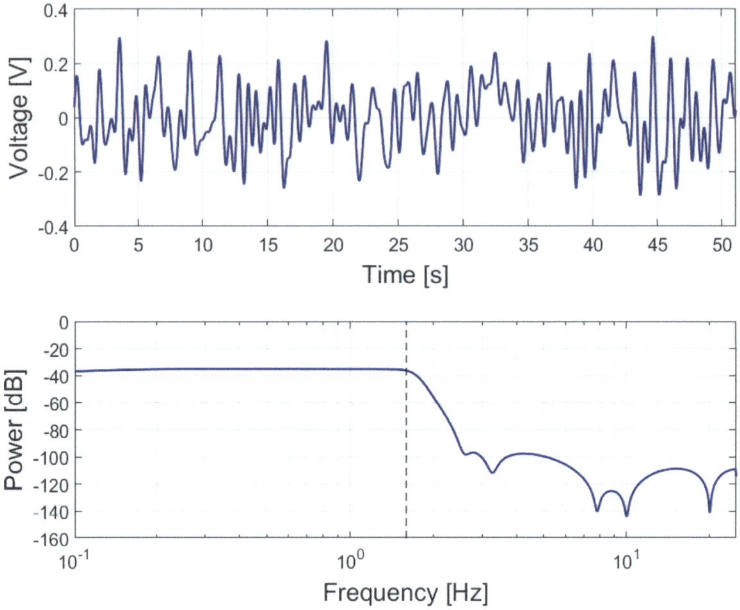

Fig. 3 The PRBS in time and frequency domain

necessary to excite the system up to a bandwidth 16 Hz, but this is unfortunately not possible due to the limitations of the steering servo. The dominant frequency range of the system is up to about 0.16 Hz. Just as the curvature of realistic trajectories is in the range below 0.2 Hz, the trajectory control should work well in this range while ensuring the robustness of the higher frequency model to uncertainties. It follows that a bandwidth of 1.6 Hz of the PRBS should be sufficient to identify the required models. The steering servo control operates with a maximum voltage signal of 0.5 V amplitude. Based on preliminary measurements, an artificial excitation of 0.3 V amplitude was required for proper excitation of the system, and therefore the amplitude of the PRBS was set to 0.3 V.

3 The Applied System ID Methods

The purpose of the ID is to mathematically describe the lateral movement of the vehicle so that appropriate control can be applied during the different autonomous functions. For the measurements, the vehicle was forced to follow a straight path in self-driving mode while the system was excited by a PRBS voltage signal applied to the steering servo.

Vehicle dynamics, including lateral dynamics, can be described by continuous-time nonlinear differential equations. The corresponding input/output relationship can be described by Volterra or Fliess functional series or by nonlinear state space models. The general form of the state space models:

$$\dot{x} = f(x, u), \qquad y = g(x, u), . \tag{1}$$

where $x \in \mathbb{R}^n$ $u \in \mathbb{R}^m$ and $y \in \mathbb{R}^p$ are the right dimensional state, input, and output vectors.

If the functions f, g are known then the system (1) can be linearized (see for example Jacobian linearization) to the

$$\dot{x}(t) = Ax(t) + Bu(t), \qquad y(t) = Cx(t) + Du(t) \tag{2}$$

state equations at a given working point.

The goal is to model from the measured data (sampled u, y), so the (2) model is discretized in time. According to the assumed supporting structure, we obtain

$$x_{t+1} = Ax_t + Bu_t, \qquad y_t = Cx_t + Du_t \tag{3}$$

discrete-time equation of state corresponding to an I/O (Input/Output) model ($d = 0$) in the time domain:

$$y_t = G(q)u_t = \sum_{k=1}^{\infty} G_k u_{t-k}, \tag{4}$$

where q is the shift operator, thus $qu_t = u_{t+1}$ and $\{G_k, k = 1, 2, \dots\}$ are the impulse response series. The discrete transfer function is using the Z-transform:

$$G(z) = \sum_{k=1}^{\infty} G_k z^{-k}. \tag{5}$$

If our system is a SISO (Single Input Single Output) system, the correct notation is $g(z) = \sum_{k=1}^{\infty} g_k z^{-k}$ where $\{g_k, k = 1, 2, \dots\}$ is the scalar impulse response series.

A linear model can also be constructed when an LPV (Linear parameter-varying) description is obtained using a physical signal (see [4, 5]):

$$x_{t+1} = A(\rho)x_t + B(\rho)u_t, \qquad y_t = C(\rho)x_t + D(\rho)u_t, \tag{6}$$

thus $y_t = G(q, \rho)u_t$, where ρ is the scheduling variable, in our case it is the vehicle speed.

In the modeling and ID of dynamic systems, environmental effects are represented by stochastic processes, so-called noise models. So we assume that the observations (measurements) are the noise-laden outputs of an LTI (Linear Time-Invariant) system.

Since in most cases we have sampled systems (data), it is obvious to use discrete-time models:

$$y_t = \sum_{k=1}^{\infty} g_k u_{t-k} + \varepsilon_t, \tag{7}$$

where ε_t is the noise process. $R_\varepsilon(k) = E\{\varepsilon_{t+k}\varepsilon_t\}$ is the covariance function where $E\{\varepsilon_t\} = 0, \forall t$; and $\Phi_\varepsilon(\omega) = \mathcal{F}\{R_\varepsilon(k)\}$ is the spectral density function.

For practical reasons, it is convenient to re-parameterize these models using a finite number of $\theta \in \mathcal{D}_\mathcal{M} \subset \mathbb{R}^d$ parameters and formulate the ID problem for this finite-dimensional set. The model is then given by

$$y_t = G(q, \theta)u_t + \varepsilon_t, \tag{8}$$

where the noise process is described by either the covariance series $R_\varepsilon(\tau)$ or the spectral density function $\Phi_\varepsilon(\omega)$. This characterization defines a $\mathcal{M}$ parameterized model set, i.e., a model structure. The element of the set $\mathcal{M}$ associated with a fixed parameter θ is denoted by $\mathcal{M}(\theta)$. Details of the parametrization issues are in references [6, 7].

Finite dimensional LTI systems can be finitely parameterized using their transfer functions. Using the appropriate shift operator, the time domain form of these models are

$$y_t = G(q, \theta)u_t + H(q, \theta)e_t, \quad e_t \sim (0, \lambda)\text{white noise} \tag{9}$$

where $H(q, \theta)$ is a stable and an inverse stable system, and $G(q, \theta)$ is stable. Then the systems G and H can be written as rational functions, like

$$G(q) = \frac{B(q)}{A(q)}, \qquad H(q) = \frac{C(q)}{D(q)}. \tag{10}$$

The resulting model structure is also called the Box–Jenkins model in the ID literature, i.e.,

$$y_t = \frac{B(q)}{A(q)}u_t + \frac{C(q)}{D(q)}e_t, \tag{11}$$

is a sum of a rational transfer function model and an ARMA (AutoRegressive Moving Average) noise model.

Based on the different assumptions made on the structure of the noise model, we can define model structures that are also relevant for the estimation of the model parameters. In the ARMA model above, if $A(q) = D(q)$ and $C(q) = 1$ then we speak of ARX (AutoRegressive with eXogeneous input), and if $C(q)/A(q) = 1$ then we speak of OE (Output Error) model structure. The parameters of the ARX model can be estimated by LS (Least Square), the parameters of the OE model by NLS (Nonlinear Least Squares), while the parameters of the general ARMA model can be estimated by ML (Maximum Likelihood) or PE (Prediction Error) (see [8]).

Define the one-step-ahead prediction error as (see [9, 10]):

$$\varepsilon(t, \theta) = H^{-1}(\theta, q)\left[y_t - G(\theta, q)u_t\right].$$

The criterion function to be minimized in the PE procedure is usually

$$V_N(\theta, Z^N) = \frac{1}{N}\sum_{t=1}^{N} \ell(\varepsilon(t, \theta)),$$

where $\ell(.)$ is a scalar-valued function.

The estimated PE parameter is

$$\hat{\theta}_N = \hat{\theta}_N(Z^N) = \arg\min_{\theta \in \mathcal{D}} V_N(\theta, Z^N),$$

where typically

$$\ell(\varepsilon(t, \theta)) = \frac{1}{2}\varepsilon^2(t, \theta)$$

choice.

The PE and ML estimates are the same for the normalization conditions on the source term.

4 ID of the Lateral Dynamics

In this chapter, we present the results obtained in identifying linear models. The whole process of estimating the structure is not described, but only the properties of the best models, based on statistical model validation, are presented as a function of the vehicle speed as a parameter, which is set constant in each experiment (measurement).

The models have been evaluated in:

- time domain (correlation function of residual series, AIC (Akaike Information Criterion), BIC (Bayesian Information Criterion),
- parameter space (estimated parameters for hypothesis T-tests $E\hat{\theta}_i = 0$, parameter covariance matrix),
- pole–zero maps, and
- the frequency domain (by comparing the APSD estimates calculated from the data with the Bode diagrams of the frequency functions calculated from the models).

The evaluation of the velocity dependence (based on the change in the position of the dominant poles was performed as the pole–zero maps). LPV ID was not applied in this phase of the study.

Based on the results of the structure estimation we get the OE model structure

$$y_\theta = \frac{b_1 q^{-1} + b_2 q^{-2}}{1 + a_1 q^{-1} + a_2 q^{-2}} u_t + \varepsilon_t \tag{12}$$

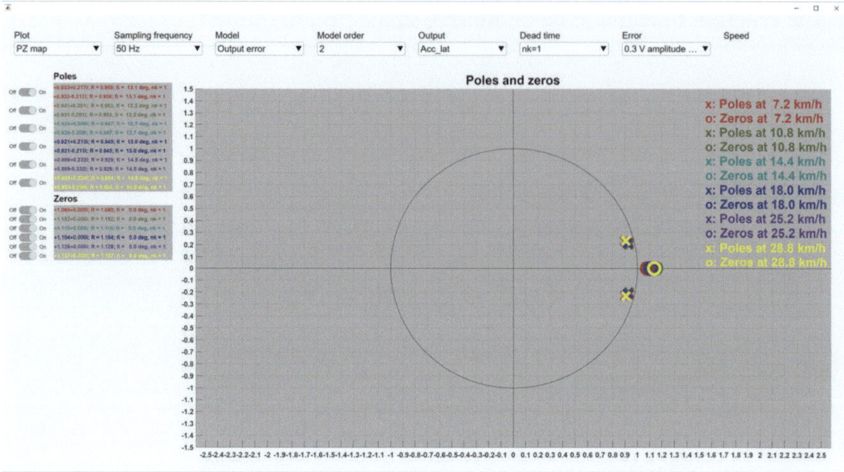

Fig. 4 The pole–zero maps of 2nd order OE models at different vehicle speeds

The OE 2nd order models were found to provide the best fit. The model parameters $\hat{\theta} = [b_1, b_2; a_1, a_2]$ were estimated using the MATLAB ID Toolbox OE procedure.

The reference signal for ID was the excited voltage signal applied to the steering servo and the output was the lateral acceleration. The sampling frequency was 50 Hz.

The Fig. 4 show the pole–zero maps of the 2nd order OE ID models and their statistics at different vehicle speeds.

The Fig. 5 shows that the system gain is minimally velocity dependent, while the BIC and FPE (Final Prediction Error) values are nearly constant (see [8] pp. 503–509).

These statistical measures were mainly used to rank the goodness of the estimates, but other measures were also taken into account (e.g., AIC, parameter scatter, T values).

5 Summary

In this paper, the ID of the lateral dynamics of a self-driving test vehicle is presented. In order to design better control structures and to take into account implementation constraints, it was necessary to define and identify a low-order model with so-called dominant poles from measured data.

The existing two-level on-board control system provided track and speed tracking in a fully autonomous mode. Excitation was provided by an artificial clutter signal added to the interfering signal. During the measurements, the vehicle was forced to

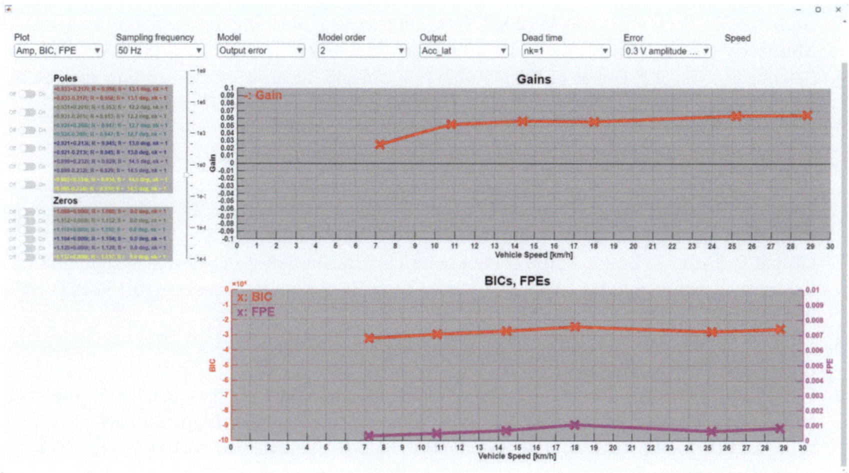

Fig. 5 The system gain, BIC and FPE statistics of 2nd order OE models at different vehicle speeds

follow a straight track in self-driving mode while the system was excited by a PRBS voltage signal. Linear models were identified from these measurements.

The ID of lateral dynamics was performed at various speed levels from measured data where the vehicle speed was constant during an experiment and the vehicle was forced to move in a straight line by the autonomous function. The maximum speed of the experiment was below $30\frac{\text{km}}{\text{h}}$, so that the nonlinearity due to the operation of the speed-dependent servo steering can be ignored.

OE ID method provided the best results for these linear models. The input of the ID was the sum of the artificial excitation voltage signal and the correction given by the controller. The output was the lateral acceleration. Based on the ID models, the second-order models gave the best results at a sampling frequency 50 Hz.

The presented methodology is suitable for ID from measurement data of other vehicles, and the reported results can be used in the maneuvering and control system design of the Nissan Leaf.

Acknowledgements The research was supported by the European Union within the framework of the National Laboratory for Autonomous Systems (RRF-2.3.1-21-2022-00002).

References

1. Berntorp K (2013) Derivation of a six degrees-of-freedom ground-vehicle model for automotive applications. Technical Reports TFRT-7627, Department of Automatic Control, Lund Institute of Technology, Lund University. https://doi.org/10.13140/2.1.1198.6723
2. Kiencke U, Nielsen L (2000) Automotive control systems for engine, driveline and vehicle. Springer. https://doi.org/10.1007/b137654

3. Bertolazzi E, Biral F, Da Lio M (2007) Real-time motion planning for multibody systems. In: Multibody system dynamics, pp 119–139. https://doi.org/10.1007/s11044-007-9037-7
4. Sename O, Gáspár P, Bokor J (2013) Robust control and linear parameter varying approaches: application to vehicle dynamics. Springer. https://doi.org/10.1007/978-3-642-36110-4
5. Gáspár P, Szabó Z, Bokor J, Németh B (2017) Robust control design for active driver assistance systems. Springer. https://doi.org/10.1007/978-3-319-46126-7
6. Bokor J, Keviczky L (1991) Parametrizations of linear stochastic-systems. Lecture notes in control and information sciences, vol 161, pp 47–65. Topics in stochastic systems: modelling, estimation and adaptive control. https://doi.org/10.1007/BFb0009298
7. Gáspár P, Szabó Z, Bokor J (2007) Parameter identification of a suspension system and road disturbance estimation. Int J Veh Syst Model Test 128–137. https://doi.org/10.1504/IJVSMT.2007.013891
8. Ljung L (1999) System Identification, theory for the user. Prentice-Hall, 2nd edn. https://doi.org/10.1109/MRA.2012.2192817
9. Caines PE (1976) Prediction error identification methods for stationary stochastic processes. IEEE Trans Autom Control 500–505. https://doi.org/10.1109/TAC.1976.1101304
10. Ljung L, van Overbeek AJM (1978) Validation of approximate models obtained from prediction error identification. In: Proceedings of the 7th IFAC congress, Helsinki, pp 1899–1980. https://doi.org/10.1016/S1474-6670(17)66164-0

Linear Parameter Varying and Reinforcement Learning Approaches for Trajectory Tracking Controller of Autonomous Vehicles

András Mihály, Van Tan Vu, Trong Tu Do, Kieu Duc Thinh, Nguyen Van Vinh, and Péter Gáspár

1 Introduction

With the outstanding development of science and technology today, autonomous vehicles have been researched and developed, bringing many conveniences to people, reducing traffic congestion, and significantly reducing traffic accidents [1]. Along with these advantages, plus most autonomous vehicles use green energy, they are greatly welcomed by drivers and encouraged and supported by the governments.

A. Mihály (✉) · P. Gáspár
Systems and Control Laboratory, Institute for Computer Science and Control (SZTAKI), Hungarian Research Network, Budapest, Hungary
e-mail: mihaly.andras@sztaki.hu

P. Gáspár
e-mail: gaspar.peter@sztaki.hu; gaspar.peter@kjk.bme.hu

P. Gáspár
Department of Control for Transportation and Vehicle Systems, Budapest University of Technology and Economics, Budapest, Hungary

V. . T. Vu · N. V. Vinh
Department of Automotive Mechanical Engineering, Faculty of Mechanical Engineering, University of Transport and Communications, Hanoi, Vietnam
e-mail: vvtan@utc.edu.vn

N. V. Vinh
e-mail: vinh181310097@lms.utc.edu.vn

T. T. Do
Mechanical and Power Engineering Faculty, Electric Power University, Hanoi, Vietnam
e-mail: tudt@epu.edu.vn

K. D. Thinh
Thuyloi University, Dong Da, Hanoi, Vietnam
e-mail: kieuducthinh@tlu.edu.vn

G. L. Conte and O. Sename (eds.), *Proceedings of 12th International Conference on Mechatronics and Control Engineering*, Lecture Notes in Mechanical Engineering,
https://doi.org/10.1007/978-981-96-6452-8_13

With this development trend, autonomous vehicles are receiving increasing investment from automobile companies and research centers around the world, turning the autonomous vehicle market into a potential market for investors.

To achieve safe mobility with little or no driver intervention, sensors, controllers, algorithms, machine learning systems, and processors must be meticulously coordinated and organized. First, information about the vehicle and surrounding environment is collected by a network of main vehicle sensors such as GPS, IMU, encoders, cameras, LIDAR, and other sensors [2–4]. This data will be processed to indicate the location, speed of the vehicle, and obstacles. After processing, a module creates a trajectory and sends it to the vehicle's controller to control speed, acceleration, steering angle, stability, and braking appropriately [5–9]. These operations are based on the vehicle's current position, a calculated sequence of references, along obstacle avoidance algorithms, thanks to which the autonomous vehicle ensures smooth, safe movement while traveling on the road. RL (Reinforcement Learning) is a branch of Machine learning, a controller integrated with RL has the ability to collect information and to learn through trial and error using feedback from its actions. Thanks to these capabilities, RL is increasingly widely applied in autonomous vehicle control to equip self-driving vehicles with higher intelligence and adaptability to the environment [10]. RL algorithms such as the Deep Neural Network (DNN) and Deep Reinforcement Learning (DRL), when equipped with autonomous vehicles, have the ability to navigate when changing lanes with predefined routes [11]. Models based on Q-learning, Q-network, and Deep Deterministic Policy Gradient (DDPG) algorithms can learn to handle unforeseen situations and need adaptability on the road [12, 13]. This adaptability is a very important and necessary factor in handling situations ranging from complex interactions during traffic to adaptation to unexpected weather conditions. This ability is continuously learned and updated while interacting with the environment to help autonomous vehicles ensure safety and efficiency when operating in different conditions and emergencies while traveling [14], obstacle avoidance [15], and longitudinal control when changing lanes.

However, the RL method applied to control the motion trajectory of autonomous vehicles still faces difficulties in some cases, for example in the case of noisy sensors. Another control method applied to autonomous vehicles is the Linear Parameter Varying (LPV) control method. LPV control theory is especially effective in dealing with complex nonlinear systems [16]. The LPV technique is capable of representing various nonlinear systems with arbitrary slow or fast changes of parameters [17]. Besides, thanks to the ability to extend some linear concepts to nonlinear systems such as $\mathcal{H}_\infty$, sensitivity shaping, and D-stability, it helps it be applied to nonlinear systems in vehicles [18]. When LPV is integrated into the autonomous vehicle controller, it is capable of autonomous driving control with time-varying speed and external disturbances [19]. Therefore, the advantages of the LPV method can overcome the RL method mentioned above, so this study aims to combine the above two methods to control the motion trajectory of autonomous vehicles. Specifically an LPV supervised RL-based trajectory tracking control approach is introduced with the purpose of ensuring vehicle stability in case of sensor interference, such as GPS signal interference. The main novelty of the present paper consists of the supervised control architecture, in which the RL-based trajectory tracking controller is supervised by a robust LPV trajectory tracking controller, with the aim to deal with possible sensor noises in the GPS system. Thus, in case of interference in the GPS signal is detected, the RL-based trajectory tracking of the autonomous vehicle is overtaken by the LPV controller. By this means, the otherwise exceptional RL controller which cannot handle disturbances and noises in a robust and stable manner is substituted by the LPV controller, which has an in-built stability margin to deal with such instances.

The paper is organized as follows: Sect. 2 describes a single-track bicycle vehicle model. Section 3 presents the LPV controller synthesis for vehicle trajectory tracking. Section 4 details the trajectory tracking RL controller for the autonomous vehicle. Section 5 demonstrates the simulation results in the CarSim simulation environment through high-speed test scenarios. Finally, concluding remarks are presented in Sect. 6.

2 Modeling of Vehicles for Trajectory Following

The bicycle model, serving as a cornerstone in vehicle simulation and dynamics analysis, provides a simplified representation of a vehicle's lateral motion and steering behavior, supporting research in high-level control design and trajectory planning, notably in the field of autonomous vehicles. In this model, which is graphed in Fig. 1, the vehicle's front and rear axles are represented as point masses, with state variables including position, orientation, velocity, and steering angle δ. Equations of motion capture the interplay of forces and moments affecting the vehicle, while a simplified steering model relates the steering angle δ to the rate of change of the yaw angle ψ. Researchers leverage the bicycle model to test and refine control algorithms, facilitating the development of autonomous vehicle systems and robotics applications. While the model has its limitations, it offers a practical and efficient framework for initial testing and analysis before transitioning to more complex and detailed vehicle models. The dynamic equations in the lateral direction are shown in Eq. (1) [20].

$$\begin{cases} J\ddot{\psi} = c_1 l_f(\delta - \beta - \dot{\psi} l_1/\dot{\xi}) - c_2 l_r(-\beta + \dot{\psi} l_2/\dot{\xi}) \\ m\dot{\xi}(\dot{\psi} + \dot{\beta}) = c_f(\delta - \beta - \dot{\psi} l_f/\dot{\xi}) + c_r(-\beta + \dot{\psi} l_r/\dot{\xi}) \\ \ddot{y}_v = \dot{\xi}(\dot{\psi} + \dot{\beta}) \end{cases} \tag{1}$$

where β is the side-slip angle, J is the yaw inertia of the vehicle, m is the mass of the complete vehicle, l_f and l_r are the distances from the front and rear axles to the Center of Gravity (CG), c_f and c_r are respectively the front and rear tire cornering stiffness, which are identified in [21]. Meanwhile $\ddot{\xi}$ and $\ddot{y}_v$ are the longitudinal and lateral accelerations, which are measured by sensors.

In the realm of autonomous vehicle research, the goal of trajectory tracking design is paramount, with a specific focus on the coordinated utilization of the coordinate axes within the bicycle model. This objective centers on the development of advanced

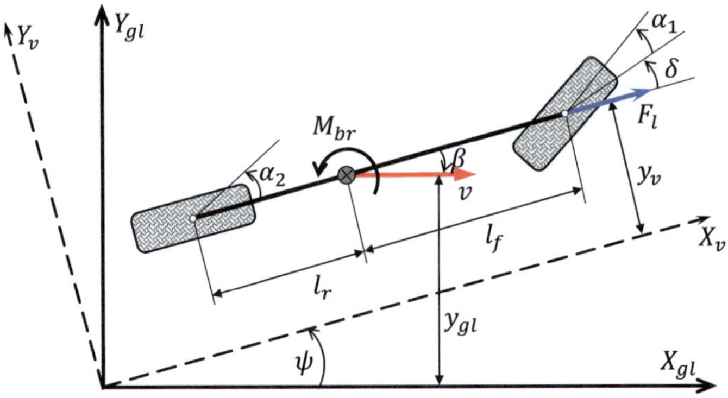

Fig. 1 Single track bicycle model

control algorithms and systems that enable autonomous vehicles to adhere to predefined trajectories while navigating complex real-world scenarios precisely. These systems must consider the vehicle's state variables, employing a coordinate system that typically encompasses longitudinal and lateral directions. The seamless coordination and control of the vehicle's motion along these axes are essential for ensuring that the vehicle accurately follows its intended path while adhering to safety standards, adapting to dynamic environments, and optimizing operational efficiency. This interdisciplinary pursuit in trajectory tracking design is pivotal for enhancing the capabilities and safety of autonomous vehicles in various contexts, spanning from path planning to real-time motion control.

$$y_{v,r} = -sin(\psi)\ x_{gl,r} + cos(\psi)\ y_{gl,r} \tag{2}$$

Therefore, in Fig. 1, the coordinate transformation conversion between the word coordinate system (X_{gl} and Y_{gl}) and the vehicle's own rotating coordinate system (X_v and Y_v) is necessary. That is, the coordinate system attached to the vehicle will rotate by the yaw angle ψ. Here, the reference road coordinates $x_{gl,r}$ and $y_{gl,r}$ are introduced for the predefined trajectory of the vehicle. Hence, the lateral position of the reference road in the coordinate system of the vehicle can be calculated in Eq. (2). Consequently, by utilizing the real-time vehicle position data obtained from the onboard GPS positioning system, it becomes possible to calculate the lateral deviation from the predetermined reference trajectory.

3 LPV Controller Design

It can be seen in Eq. (1) that the vehicle model depends on the forward velocity $\dot{\xi}$ and the inverse of the forward velocity $\frac{1}{\dot{\xi}}$. Moreover, when the vehicle is in motion, the forward velocity is one of the constantly changing parameters, and it depends on the driver and the motion condition of the vehicle. Here, the forward velocity $\dot{\xi}$ is chosen as a scheduling parameter. The dynamic equation (1) can be rewritten in the LPV state-space representation as in Eq. (3) with the varying parameter $\rho = \dot{\xi}$.

$$\dot{x} = A(\rho)x + B_1(\rho)w + B_2(\rho)u \tag{3}$$

where the matrices $A(\rho)$, $B_1(\rho)$, and $B_2(\rho)$ are expressed in the following form:

$$\begin{cases} A(\rho) = A_0 + A_1\rho + A_2\frac{1}{\rho} \\ B_1(\rho) = B_{10} + B_{11}\rho \\ B_2(\rho) = B_{20} + B_{21}\rho \end{cases} \tag{4}$$

The state vector of the system is selected as $x = \left[\dot{\psi}\ \beta\ \dot{y}_v\ y_v\right]^T$ includes the yawrate, side-slip angle, lateral velocity and position of the vehicle.

In Fig. 2, the given H_∞/LPV control structure includes the nominal model $G(\rho)$, the controller $K(\rho)$, the performance output z, the control input u, the measured output y, the measurement noise n.

According to Fig. 2, the concatenation of the nonlinear model (3) with the performance weighting functions has a partitioned representation in the following way:

$$\begin{bmatrix} \dot{x}(t) \\ z(t) \\ y(t) \end{bmatrix} = \begin{bmatrix} A(\rho) & B_1(\rho) & B_2(\rho) \\ C_1(\rho) & D_{11}(\rho) & D_{12}(\rho) \\ C_2(\rho) & D_{21}(\rho) & D_{22}(\rho) \end{bmatrix} \begin{bmatrix} x(t) \\ w(t) \\ u(t) \end{bmatrix} \tag{5}$$

with the exogenous input $w(t) = \left[\,R\ n\,\right]^T$, the control input $u(t) = \left[\,\delta\,\right]^T$, the measured output vector $y(t) = \left[\,\dot{\xi}\ y_v\,\right]^T$.

The performance output is defined as $z(t) = \left[z_1\ z_2\right]^T$. In which, $z_1 = |y_{v,r} - y_v|$ is the lateral deviation from the designed path. The aim of the control problem is

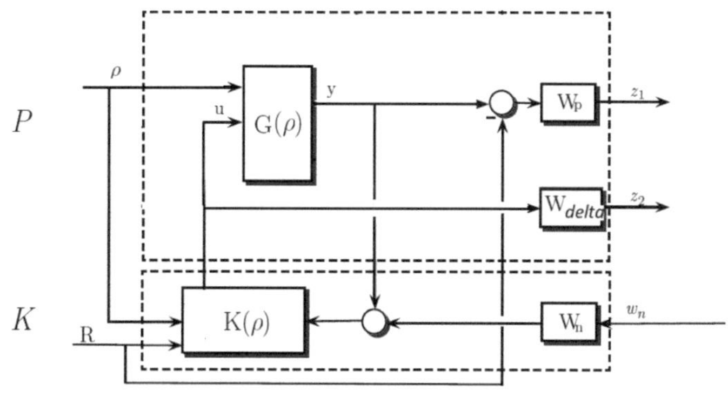

Fig. 2 Closed-loop interconnection structure

to minimize the value of z_1, i.e. $z_1 \rightarrow 0$ (optimization criterion). Meanwhile, the second output performance is selected as follows: $z_2 = \left[\delta\right]^T$. The main goal of this idea is to reduce the steering angle while ensuring appropriate trajectory tracking, that is, to avoid the saturation of the actuator. The choice of the performance output is very important and affects the effectiveness of the controller design.

The weighting functions W_n and W_δ represent respectively the sensor noises of the system (in this case the noisy GPS positioning signal), the activation of steering intervention, and avoiding actuator saturation. With T_1 and λ are designed parameters, these weighting functions are chosen as first-order proportional form:

$$\lambda \frac{1}{T_1 s + 1} \tag{6}$$

The weighting function W_p represents the performance signals ensuring the velocity and path tracking of the autonomous vehicle, and is chosen in the following form:

$$W_p = \lambda \frac{\alpha_2 s^2 + \alpha_1 s + 1}{T_1 s^2 + T_2 s + 1}, \tag{7}$$

where $T_{1,2}$, $\alpha_{1,2}$ and λ, are designed parameters. Note, that these design parameters are selected for shaping the weighting functions in a manner to realize the expected behavior of the closed-loop system. For example, W_p weighting function can be considered as a penalty function, thus the shaping is designed to suppress the performance signals in the desired frequency range. On the other hand, W_δ is shaped in a manner to allow a bigger steering angle for smaller frequencies and to avoid saturation of the steering system. The selection of design parameters are chosen in an iterative manner, adjusting necessary parameters after the analysis of the closed-loop system.

The LPV controller $K(\rho)$ minimizes the induced $\mathcal{L}_2$ norm of the closed-loop LPV system $\sum_{CL}(\rho) = LFT(\sum(\rho), K(\rho))$, with zero initial conditions, i.e.:

$$\| \sum_{CL}(\rho) \|_{2\rightarrow 2} = \sup_{\substack{\rho \in \mathcal{P} \\ \bar{\nu} \leq \dot{\rho} \leq \underline{\nu}}} \sup_{\substack{w \in \mathcal{L}_2 \\ \|w\|_2 \neq 0}} \frac{\| z \|_2}{\| w \|_2} \tag{8}$$

The existence of a controller that solves the parameter-dependent LPV γ-performance problem can be expressed as the feasibility of a set of Linear Matrix Inequalities (LMIs), which can be solved numerically [22–24].

The control goal is to find an LPV controller $K(\rho)$, which minimizes the induced $\mathcal{L}_2$ norm of the closed-loop LPV system $\sum_{CL}(\rho) = LFT(\sum(\rho), K(\rho))$, with zero initial conditions. The induced $\mathcal{L}_2$ norm from the disturbance ω to the performances z remains smaller than a predefined γ, as detailed in [25].

4 Design of LPV Supervised Reinforcement Learning Control

A deep Q-learning network (DQN) agent has been applied, while the reinforcement learning environment consists of the two-wheeled bicycle model, which represents the ego vehicle dynamics. The goal of the training is to realize trajectory tracking for the autonomous vehicle by applying an appropriate steering angle. A $Ts = 0.02$ s sample time has been used for the training process, while the output of the trajectory tracking control is the steering angle of the vehicle. In order to consider the physical limitations of the ego car, the steering angle has to be in the range of $[-37 \quad 37]$ deg. The trajectory of the road is given by the simulation road presented in Sect. 5. The initial condition for the lateral deviation is set to 0.2 m, while the relative yaw angle -0.1 rad has been chosen.

The training model for the ego vehicle has been built in the following manner:

- The steering action from the agent to the environment is selected from $-37°$ to $37°$.
- Observations from the environment are defined as follows:
  - Lateral deviation: e_y
  - Relative yaw angle: e_ψ
  - Derivative of lateral deviation: $\dot{e_y}$
  - Derivative of relative yaw angle: $\dot{e_\psi}$
  - Integral of lateral deviation: $\int e_y$
  - Integral of relative yaw angle: $\int e_\psi$

The calculated reward r_t given at each time step t is chosen as $r_t = 20e_y^2 + 0{e_\psi}^2 + 20u^2 + 0\dot{e_y}^2 + 0\dot{e_\psi}^2$, where the episode reward is the cumulative value of r_t. An example of the training process is shown in Fig. 3.

The aim of the LPV supervised control architecture is to guarantee the stability of the vehicle using the RL controller even in the case of disturbances or sensor noises, which are assumed to be detected as detailed in [26]. In normal operating conditions, the vehicle steering system is operated by the RL agent, while in case of detected sensor noise, the robust LPV controller overtakes the control action, as depicted in Fig. 4. Note, that the two designed controllers operate in parallel, with the LPV controller designing a steering angle δ_{LPV}, whereas the RL agent calculates δ_{RL}. The following simple decision logic is applied for choosing the steering input of the autonomous vehicle:

$$\delta = \begin{cases} \delta_{RL}, & \text{without noise} \\ \delta_{LPV}, & \text{with noise} \end{cases} \tag{9}$$

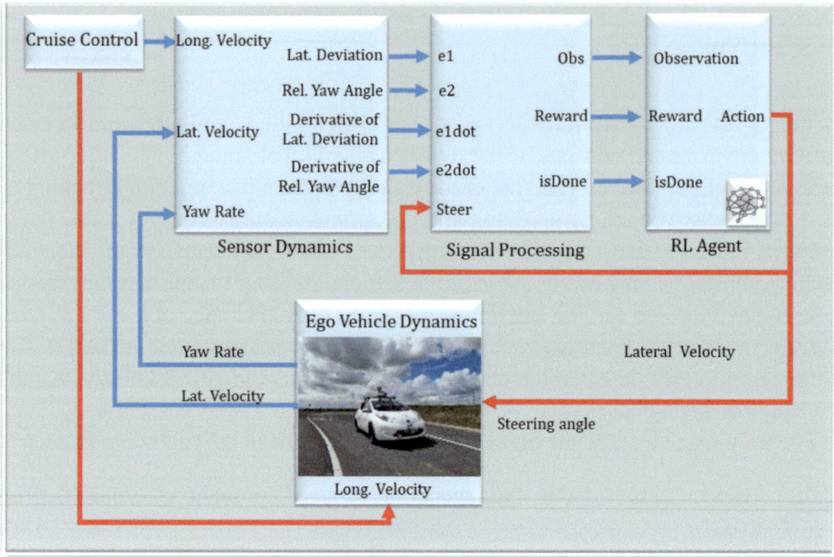

Fig. 3 Reinforcement learning training process

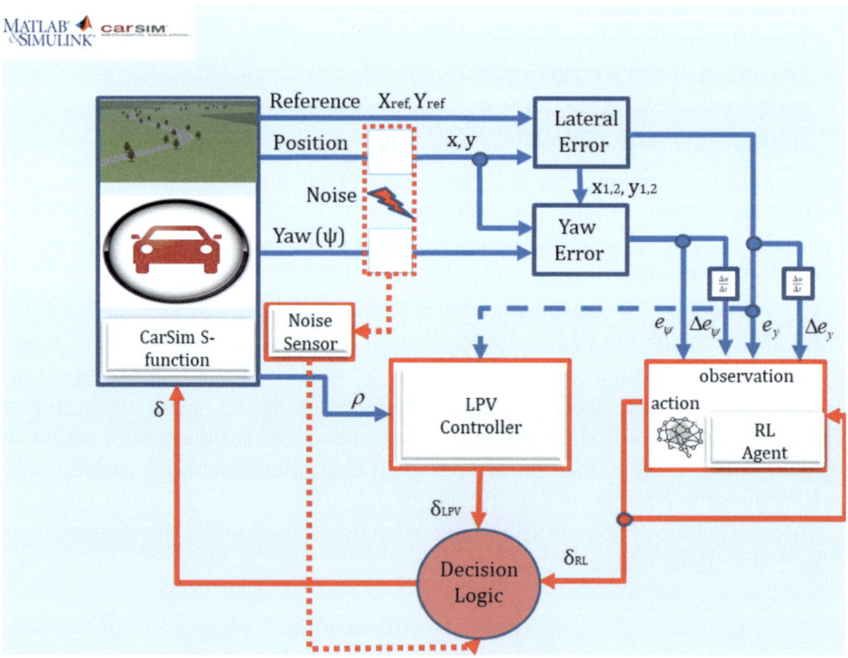

Fig. 4 RL agent supervised controller

5 Simulation Results

To demonstrate the operation of the introduced reinforcement learning-based supervised control design, real-life simulation scenarios have been tested with and without the designed supervised control method in the absence or presence of sensor noise associated with the GPS positioning system. The geometric and dynamic parameters of the chosen simulation vehicle are illustrated in Table 1. Note, that the same parameter values have been applied for the LPV control design described in Sect. 3 and for the RL training process detailed in Sect. 4.

The simulations have been evaluated with a constant velocity of 100 km/h on a curvy highway road given in CarSim environment as depicted in Fig. 5.

Table 1 Autonomous vehicle parameters

Parameter	Value	Unit
Vehicle mass (m)	1742	kg
Yaw moment of inertia (J)	1523	kgm^2
Distance from C.G to front axle (l_1)	1.161	m
Distance from C.G to rear axle (l_2)	1.539	m
Tread front (b_f)	1.54	m
Vehicle mass Tread rear (b_r)	1.535	m
Height of COG (h_{COG})	0.438	m
Cornering stiffness front (c_1)	70	kN/rad
Cornering stiffness rear (c_2)	55	kN/rad
Aerodynamic drag co-efficient (c_w)	0.3431	–
Vehicle mass Front contact surface (A)	1.6	m^2

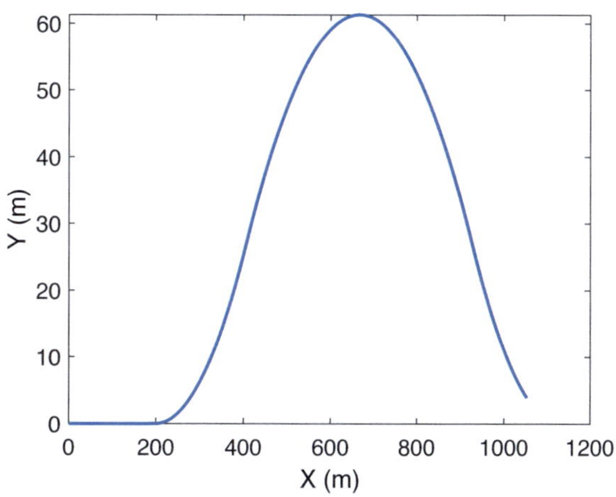

Fig. 5 Highway route X-Y plane

5.1 Reinforcement Learning Control

During the base simulation case, the steering of the autonomous vehicle is operated by the trained RL agent. Here, sensor noise of the GPS system is not assumed, thus in this fault-free case trajectory tracking is performed as expected with minimal lateral error, see Fig. 6b. Here, the maximal lateral error never exceeds 2 cm. The corresponding steering wheel angle calculated by the RL controller is shown in Fig. 6a. Note, that both the yawrate depicted in Fig. 6c and the lateral velocity shown in Fig. 6d remain in a comfortable range for the passengers of the automated vehicle.

5.2 Reinforcement Learning Control with Sensor Noise

In the next simulation case, a sensor noise for the navigation system has been added, as depicted in Fig. 7. Thus, a band-limited white noise signal with a maximal amplitude of 0.3 m has been introduced to the position of the vehicle between 5–11 s.

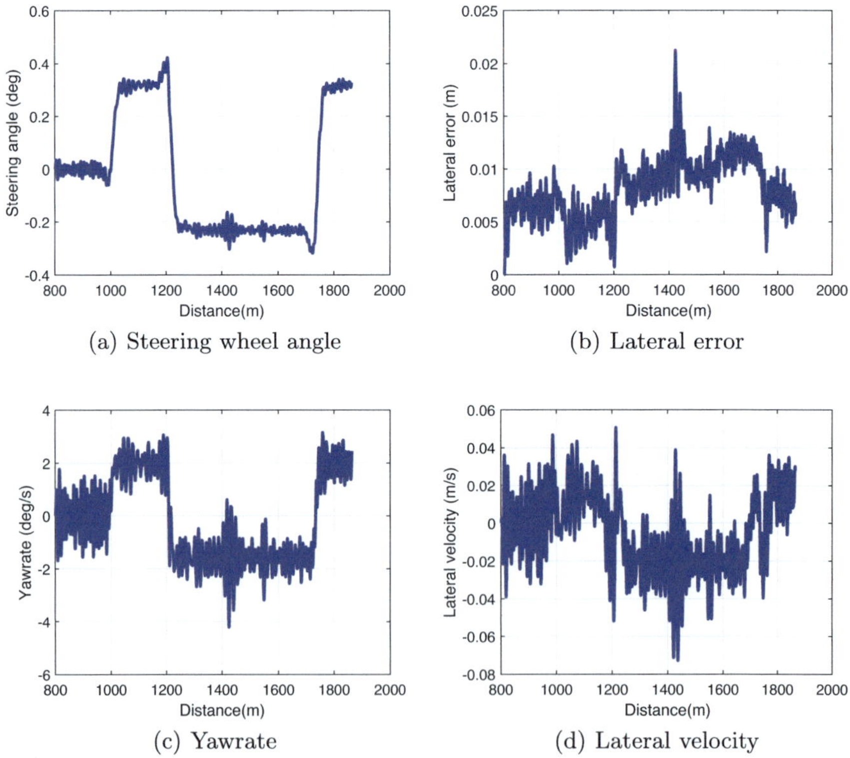

(a) Steering wheel angle (b) Lateral error (c) Yawrate (d) Lateral velocity

Fig. 6 Reinforcement learning without noise

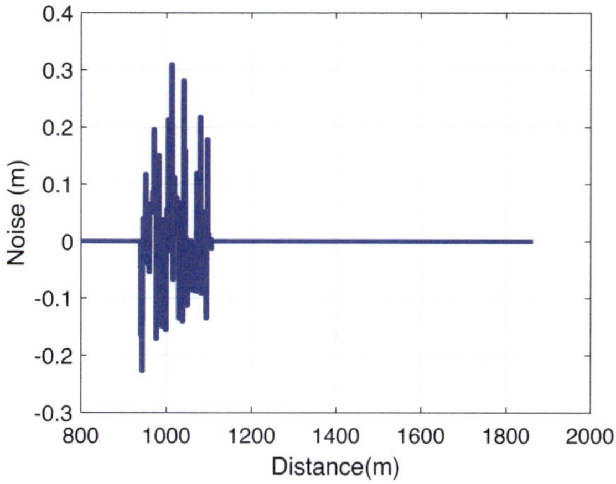

Fig. 7 Sensor noise

As it is shown in Fig. 8, the defined GPS noise destabilizes the autonomous vehicle. The lateral error from the designed path depicted in Fig. 8b quickly exceeds 2 m, while the steering input of the autonomous vehicle begins to oscillate, see Fig. 8a. As a result, both the yawrate depicted in Fig. 8c and the lateral velocity sown in Fig. 8d exceeds critical levels as the autonomous vehicle skids out of the lane.

5.3 Supervised RL Control with Sensor Noise

Lastly, the simulation scenario has been selected where the autonomous vehicle uses the LPV supervised control architecture described in Fig. 4. Thus, in case the sensor noise of the GPS is detected, the trajectory tracking control of the autonomous vehicle is taken over by the LPV controller from the RL agent. Hence, although the lateral error of the vehicle shown in Fig. 9b increases compared to fault-free case, it does not exceed 0.25 m, which is acceptable for high-speed cornering with noisy GPS signals, as well as the slightly oscillating steering wheel angle, see Fig. 9a. Note, that both the yawrate depicted in Fig. 9c and the lateral velocity shown in Fig. 9d increases as the sensor noise appears, however, quickly decreases as the GPS signal is normalized, thus the stability of the autonomous vehicle can be preserved.

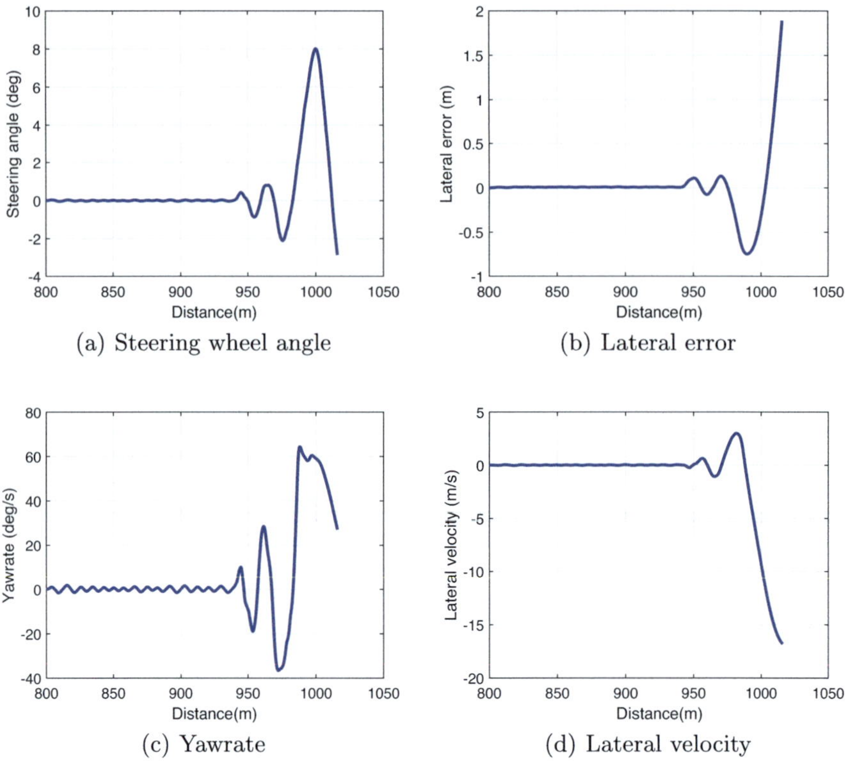

(a) Steering wheel angle (b) Lateral error

(c) Yawrate (d) Lateral velocity

Fig. 8 Reinforcement learning with noise

6 Conclusion

The paper proposed a supervised control method in which an LPV controller supervises the trained reinforcement learning agent to realize stable trajectory tracking of the autonomous vehicle even under the effect of sensor noises. Thus, in case a faulty or noisy GPS signal is detected, the operation of the steering is governed by the LPV controller, while in normal conditions by the reinforcement learning agent. The proposed supervised control method has been validated by simulations performed in the CarSim environment. In future work, sensor noise detection and controller selection and switching methods will be further developed.

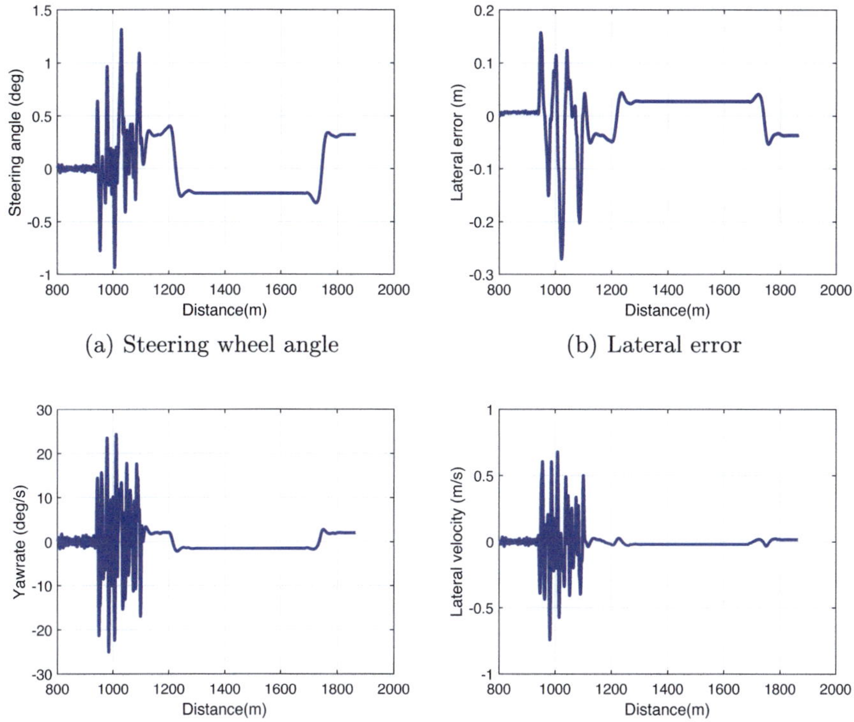

(a) Steering wheel angle (b) Lateral error

(c) Yawrate (d) Lateral velocity

Fig. 9 LPV supervised reinforcement learning with noise

Acknowledgements The research was partially supported by the European Union within the framework of the National Laboratory for Autonomous Systems (RRF-2.3.1-21-2022- 00002). The research was also supported by the National Research, Development, and Innovation Office through the project "Cooperative emergency trajectory design for connected autonomous vehicles" (NKFIH: 2019-2.1.12-TÉT_VN).

References

1. Yurtsever E, Lambert J, Carballo A, Takeda K (2020) A survey of autonomous driving: common practices and emerging technologies. IEEE Access 8:58 443–58 469
2. Min H, Wu X, Cheng C, Zhao X (2019) Kinematic and dynamic vehicle model-assisted global positioning method for autonomous vehicles with low-cost gps/camera/in-vehicle sensors. Sensors 19(24):5430
3. Sukkarieh S, Nebot EM, Durrant-Whyte HF (1999) A high integrity IMU/GPS navigation loop for autonomous land vehicle applications. IEEE Trans Robot Autom 15(3):572–578
4. Van Wyk F, Wang Y, Khojandi A, Masoud N (2019) Real-time sensor anomaly detection and identification in automated vehicles. IEEE Trans Intell Transp Syst 21(3):1264–1276

5. Falcone P, Borrelli F, Asgari J, Tseng HE, Hrovat D (2007) Predictive active steering control for autonomous vehicle systems. IEEE Trans Control Syst Technol 15(3):566–580
6. He X, Liu Y, Lv C, Ji X, Liu Y (2019) Emergency steering control of autonomous vehicle for collision avoidance and stabilisation. Veh Syst Dyn 57(8):1163–1187
7. Kang CM, Lee S-H, Chung CC (2018) Discrete-time LPV h_2 observer with nonlinear bounded varying parameter and its application to the vehicle state observer. IEEE Trans Industr Electron 65(11):8768–8777
8. Corno M, Panzani G, Roselli F, Giorelli M, Azzolini D, Savaresi SM (2020) An LPV approach to autonomous vehicle path tracking in the presence of steering actuation nonlinearities. IEEE Trans Control Syst Technol 29(4):1766–1774
9. Ren T, Wang B, GU G, Chen X (2023) Adaptive velocity and acceleration control of autonomous vehicle systems
10. Elallid BB, Benamar N, Hafid AS, Rachidi T, Mrani N (2022) A comprehensive survey on the application of deep and reinforcement learning approaches in autonomous driving. J. King Saud Univ.-Comput. Inf. Sci. 34(9):7366–7390
11. Liao J, Liu T, Tang X, Mu X, Huang B, Cao D (2020) Decision-making strategy on highway for autonomous vehicles using deep reinforcement learning. IEEE Access 8:177 804–177 814
12. Ye Y, Zhang X, Sun J (2019) Automated vehicles behavior decision making using deep reinforcement learning and high-fidelity simulation environment. Transp Res Part C: Emerg Technol 107:155–170
13. Zhang H, Xu J, Qiu J (2022) An automatic driving control method based on deep deterministic policy gradient. Wirel Commun Mob Comput 2022:1–9
14. Wang P, Chan C-Y, de La Fortelle A (2018) A reinforcement learning based approach for automated lane change maneuvers. In: IEEE Intelligent vehicles symposium (IV). IEEE, pp 1379–1384
15. Gong C, Li Z, Lu C, Gong J, Hu F (2019) Comparative study on transferable driver behavior learning methods in the lane-changing scenario. In: IEEE intelligent transportation systems conference (ITSC). IEEE, pp 3999–4005
16. Mohammadpour J, Scherer CW (2012) Control of linear parameter varying systems with applications. Springer Science & Business Media
17. Briat C (2014) Linear parameter-varying and time-delay systems. Anal Observat Filter Control 3:5–7
18. Sename O, Gaspar P, Bokor J (2013) Robust control and linear parameter varying approaches: application to vehicle dynamics, vol 437. Springer
19. Li P, Nguyen A-T, Du H, Wang Y, Zhang H (2021) Polytopic LPV approaches for intelligent automotive systems: state of the art and future challenges. Mech Syst Signal Process 161:107931
20. Kiencke U, Nielsen L (2005) Automotive control systems. Springer, Berlin, Heidelberg
21. Szabó B (2012) Vehicle test based validation of a tire brush model using an optical velocity sensor. Periodica Polytechnica 40:33–38
22. Gaspar P, Bokor J, Szaszi I (2004) The design of a combined control structure to prevent the rollover of heavy vehicles. Eur J Control 10(2):148–162
23. Gaspar P, Szabo Z, Bokor J (2005) Prediction based combined control to prevent the rollover of heavy vehicles. In: Proceedings of the 13th mediterranean conference on control and automation, Limassol, Cyprus
24. Gáspár P, Szabó Z, Bokor J (2005) The design of an integrated control system in heavy vehicles based on an LPV method. In: Proceedings of the 44th IEEE conference on decision and control. IEEE, pp 6722–6727
25. Wu F, Yang XH, Packard A, Becker G (1996) Induced l^2-norm control for LPV systems with bounded parameter variation rates. Int J Nonlinear Robust Control 6:983–998
26. Brás S, Rosa P, Silvestre C, Oliveira P (2015) Fault detection and isolation in inertial measurement units based on bounding sets. IEEE Trans Autom Control 60:1933–1938

Safe Control Framework of Autonomous Vehicles for Collision Prevention

Attila Lelkó, Balázs Németh, and Péter Gáspár

1 Introduction and Motivation

Advances in computer technology resulted in cheap and fast computers and the spread of data-based methods, machine learning, and artificial neural networks. These algorithms provide promising results in most engineering applications, i.e., transportation engineering and control theory. Various performance specifications can be defined for autonomous vehicle control systems. Usually, there are primary performance specifications, which due to safety reasons must be kept, such as guaranteeing stable vehicle motion, reliability, or keeping different traffic regulations. Moreover, several further non-safety performance requirements can be defined, which have lower priority, e.g., providing passenger comfort, achieving economic motion, minimization of traveling times, etc. Lots of performance criteria together with the complex vehicle environment lead to challenging problems for robust and optimal control design methods.

Learning-based approaches are generally applied in data-driven modeling and control solutions for highly nonlinear problems. The effectiveness of such methods over traditional algorithms is investigated in several works, i.e., [1–3]. Modern learning-based solutions usually result in complex nonlinear structures and their performance cannot be guaranteed through traditional methods. There are some studies, however, that investigate possible methods for analysis by linearizing in a large number of operating points and by forming a polytopic linear system [4]. Different approaches can be found in [5, 6], and [7] where the verification is carried out in special cases but not generally. Nevertheless, it is a current challenge to find control structures, with which safe and high-performance operation for such complex systems can be achieved [8]. The problem is solved by using a supervisor which is able

A. Lelkó (✉) · B. Németh · P. Gáspár
Hungarian Research Network, Institute for Computer Science and Control (HUN-REN SZTAKI), Budapest, Hungary
e-mail: attila.lelko@sztaki.hun-ren.hu

G. L. Conte and O. Sename (eds.), *Proceedings of 12th International Conference on Mechatronics and Control Engineering*, Lecture Notes in Mechanical Engineering,
https://doi.org/10.1007/978-981-96-6452-8_14

to combine classical robust control methods with the learning-based elements [9]. The work of this paper uses the latter method to achieve high-performance robust control for an F1TENTH [10] type of small-scaled indoor test vehicle.

Different solutions for motion control of small-scaled vehicles are presented in [11, 12], however, both solutions are based on Model Predictive Control instead of purely learning-based neural networks. The main contribution of this paper includes a reinforcement learning-based vehicle control solution, which is augmented with a supervisor to be able to handle scenarios that may have not been present in the training set and prevent collisions with identified objects on the road.

Section 2 presents an overview of the supervised control framework, Sect. 3 details the training process of the reinforcement learning-based control agent, and Sect. 4 introduces the main aspect of the robust control design. The optimization task to achieve safe motion and collision prevention is formulated in Sect. 5. The operation of the control system is illustrated in Sect. 6. Finally, the work is concluded in Sect. 7.

2 Overview of the Supervised Control Framework

In this overview, the main idea of the proposed supervised control framework is introduced. The scheme of the system is illustrated in Fig. 1. The control architecture contains three main components. The nonlinear control agent can be any kind of nonlinear controller, typically a system that is trained using a large amount of measurement data. This controller provides high-performance operation, but the performance, stability, and robustness cannot be guaranteed theoretically. The robust controller operates parallel with a more simple structure, typically a robust linear or

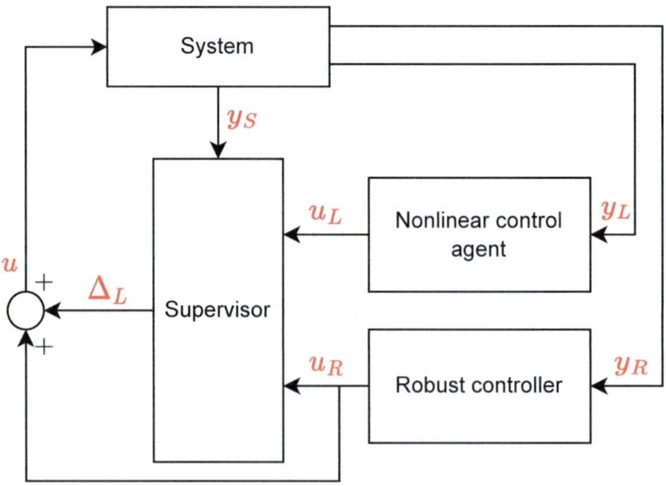

Fig. 1 Block diagram of the complete supervised control system

robust LPV controller with theoretical guarantees on stability. The third component is the supervisor which combines the advantages of the other two controllers.

The output of the supervisor is Δ_L signal, resulting in the actual control signal of

$$u = u_R + \Delta_L, \tag{1}$$

where u_R is the output of the robust controller. The dimensions of Δ_L correspond to the number of control input signal ($\Delta_L = [\Delta_{L_1}, \ \Delta_{L_2}, \ \dots]^{\mathrm{T}}$) and

$$\Delta_{L_i} \in [\Delta_{L_i,\min}, \Delta_{L_i,\max}] \quad i = 1, 2 \dots r \tag{2}$$

is a bounded addition to u_R, where r is the number of control inputs. The supervisor calculates Δ_L based on u_L, u_R and measurements on the plant y_S. The selection of Δ_L leads to a constrained optimization task, where the cost function is $\|u - u_L\|_2^2$, i.e., the actual control signal must be as close to u_L as possible to achieve high-performance operation, and the constraint ensures stability. Such constraints are, e.g., (2) to guarantee that the addition to u_R is bounded.

The object avoidance of the control system is achieved in the supervisor as well. It uses the plant model and measurements (y_S) to predict the vehicle trajectory and to recognize potential accident hazards (e.g., leaving the track or crashing). The result of the prediction is then incorporated into the optimization process through the objective function and the constraints.

The supervised control framework can achieve performance as close as possible to the nonlinear controller used in the control architecture while maintaining the robustness and stability guarantees of a robust controller. The design consideration of the components in the framework is detailed in the further sections.

3 Training of the RL-Based Vehicle Control System

The nonlinear controller in this paper is realized by a reinforcement learning-based control agent. The goal of the controller is to find the optimal steering angle (δ) and reference velocity (v_{ref}) to achieve minimal lap time without leaving the track. In this section, the selection of the most important design parameters is presented.

3.1 Vehicle Model for Training Purposes

The dynamic model used in the training environment is based on the F1TENTH vehicle. A dynamic two-wheeled bicycle model was used to achieve an accurate description without requiring large computational effort by using a large degree of freedom vehicle model. This can decrease training times and the sample time of the supervisor.

The equations of motion in the vehicle coordinate system are formulated as

$$m\dot{v}_x = F_{\text{drive}} + m v_y \dot{\psi}, \tag{3a}$$

$$m\dot{v}_y = F_{\text{Lat,F}} + F_{\text{Lat,R}} - m v_x \dot{\psi}, \tag{3b}$$

$$\theta_z \ddot{\psi} = F_{\text{Lat,F}} l_{\text{F}} - F_{\text{Lat,R}} l_{\text{R}}, \tag{3c}$$

where F_{drive} is the driving force calculated based on the reference velocity by a PID controller, $F_{\text{Lat,F}}$ and $F_{\text{Lat,R}}$ are the front and rear lateral tire forces, l_{F} and l_{R} determines the position of the center of gravity from the front and rear axle.

The lateral vehicle dynamics is enhanced with the Pacejka tire-force model [13]. The tire force is calculated as

$$F_{\text{Lat}} = -F_z \cdot \mathcal{D} \sin(\mathcal{C} \arctan(\mathcal{B}\alpha - \mathcal{E}(\mathcal{B}\alpha - \arctan(\mathcal{B}\alpha)))), \tag{4}$$

where F_z is the vertical force acting on a specific tire, $\mathcal{B}$, $\mathcal{C}$, $\mathcal{D}$ and $\mathcal{E}$ are model parameters, and α is tire lateral slip angle. The tire forces are calculated using this characteristic in the training environment.

3.2 *Training of the Reinforcement Learning-Based Control Agent*

The control agent uses a feedforward neural network with two outputs corresponding to the reference steering angle and velocity of the vehicle. The training of the neural network was performed in a reinforcement learning framework using the Proximal Policy Optimization (PPO) algorithm [14]. During the training process, the motion of the vehicle model along various tracks is simulated. These tracks were based on real-life down-scaled racing tracks or constructed from simple primitives, i.e., line and arc segments.

The inputs of the neural network consist of the local path information in the form of coordinates of N points around the vehicle. In addition, the longitudinal vehicle speed and the steering angle corresponding to the previous step are used to form a recurrent structure. In the case of the F1TENTH vehicle, $N = 5$ points provide adequate information about the surrounding track, resulting in the input vector of

$$\begin{bmatrix} X_{\text{v1}} & Y_{\text{v1}} & \dots & X_{\text{v5}} & Y_{\text{v5}} & v_{x,t-1} & \delta_{t-1} \end{bmatrix}^{\text{T}}, \tag{5}$$

where the track point coordinates are measured in the frame of reference of the vehicle, $v_{x,t-1}$ is the longitudinal velocity, and δ_{t-1} is the steering angle from the previous step.

The performance of the vehicle is measured using the following parametric reward function:

$$R(s, a) = -Ax^2_{\text{Lat,err}} - B\delta^2 + \Delta p, \tag{6}$$

where $x_{\text{Lat,err}}$ is the lateral path tracking error, Δp is the progress along the centerline since the last time step using an appropriate metric. The training algorithm aims to maximize the reward collected in the simulated episodes.

The driving behavior of the agents is significantly influenced by the weights of the reward function. Parameter A corresponds to the path-tracking capabilities of the agents, while B can decrease the control energy and oscillations of the vehicle by minimizing the steering interventions. If these weights choose to be smaller, the dominant term is the progress along the track, and the optimal solution tends to be faster lap completion.

4 The Robust Lateral Control and Safe Velocity Profile

The design of the robust control is based on the method presented in this section. The actual control signal (u) may differ from the robust control signal (u_R) by an additional Δ_L value. This difference can be handled in the controller design as a bounded input disturbance on the robust control signal, since $\Delta_L \in [\Delta_{L,\min}, \Delta_{L,\max}]$. Thus, from the viewpoint of the robust controller, the internal structures of the supervisor and the learning-based control agent are not considered. These are represented by the disturbance term Δ_L.

The robust control consists of two main parts such as the robust lateral controller and the safe velocity profile of the vehicle. In the case of the robust lateral controller, several methods are available (e.g., $\mathcal{H}_\infty$ or robust LPV methods), with which theoretical guarantees against bounded disturbances can be achieved. The design is based on the linearized version of the vehicle motion model in (3b)–(3c).

The safe longitudinal control is achieved through the determination of a safe velocity profile. This reference velocity profile is calculated based on the curvature of the track and the physical limits on the tire traction. Considering the maximum allowable acceleration leads to the inequality

$$a_{\max} \geq \|a\|_2 = \sqrt{a_x^2 + a_y^2} = \sqrt{a_x^2 + v^4\kappa^2}, \tag{7}$$

where a denotes the acceleration and v the velocity of the vehicle, and κ is the road curvature. By rearranging (7), the velocity with which the acceleration of the vehicle can be kept within limits while following the reference path can be expressed as:

$$v \leq \frac{1}{\sqrt{\kappa}}\sqrt[4]{a^2_{\max} - a^2_x}. \tag{8}$$

In straight sections, the curvature of the track is 0, which results in infinite reference velocity. Thus, v_{ref} must be limited, especially in case of low curvature values, such as $v_{\text{ref}} \leq v_{\max}$, where $v_{\max}$ is the maximum velocity limit on the given road section.

5 Supervisory Optimization to Prevent Collisions

In this section, the optimization task performed by the supervisor is detailed to combine the reinforcement learning-based control agent and the robust controller. The goal of the safe control framework is to provide u, which is close to u_L, but stable motion is guaranteed. In this application $r = 2$, the output of the supervisor is $\Delta_L = [\Delta_{L_1}, \Delta_{L_2}]^{\mathrm{T}}$. The vehicle motion can be predicted on a time horizon T using the vehicle model (3) for different K length control input sequences $\Delta_L(1), \Delta_L(2), \ldots \Delta_L(K)$. With this notation $\Delta_{L_i}(k)$ corresponds to the ith control input of the kth element in the prediction horizon. Sequences that result in the violation of the maximum path tracking error or object avoidance must be excluded from the actual control signal.

If the coordinates of the track centerline around the vehicle are known the maximum lateral error within the prediction horizon can be calculated ($e_{\text{Lat},T}$). Similarly, the minimal distance to any identified object can be estimated with the model-based prediction within the time horizon ($d_{\text{obj},T}$). The limit of the lateral error ($e_{\max}$) and the safe distance (d_{safe}) are considered as constraints during the optimization.

The objective function within the optimization considers the distance between the actual control signal and the signal of the RL agent. For practical reasons, the safe distance is included in the cost function and through the constraints as well. This resulted in a more robust algorithm altogether.

Finally, the optimization problem with nonlinear constraints for object avoidance is formulated as

$$\underset{\Delta_L}{\operatorname{argmin}} \sum_{k=1}^{K} \left(\|u_R + \Delta_L(k) - u_L\|_2^2\right) + \max\left(\frac{1}{d_{\text{obj},T}}, \frac{1}{d_{\text{safe}}}\right) \tag{9a}$$

subject to

$$\Delta_{L_i}(k) \in [\Delta_{L_i,min}; \Delta_{L_i,max}], \quad i = 1, 2, \quad k = 1, 2, \ldots K \tag{9b}$$

$$e_{\text{Lat},T} \leq e_{\max}, \tag{9c}$$

$$d_{\text{obj},T} \geq d_{\text{safe}}. \tag{9d}$$

Remark that $\Delta_{L_i,min}$ and $\Delta_{L_i,max}$ are design parameters, which significantly influence the operation of the closed-loop. The selection of larger intervals results in a more conservative robust controller and the domination of the RL agent, while in the case of tighter intervals, the robust control signal will be used most of the time.

6 Demonstration of the Control System Through Simulation Example

The effectiveness of the proposed control system through simulations is demonstrated. The RL agent was trained on various tracks using the reward function

$$R(s, a) = -5x_{\text{Lat,err}}^2 - 0.3\delta^2 + \Delta p, \tag{10}$$

which can provide accurate path tracking and control input limitation. The robust controller is $\mathcal{H}_\infty$ controller, the input disturbance is limited by $\Delta_{L_1} \in [-0.3, 0.3]$ and $\Delta_{L_1} \in [-1, 0.2]$ corresponding to the steering angle and the safe velocity profile. The

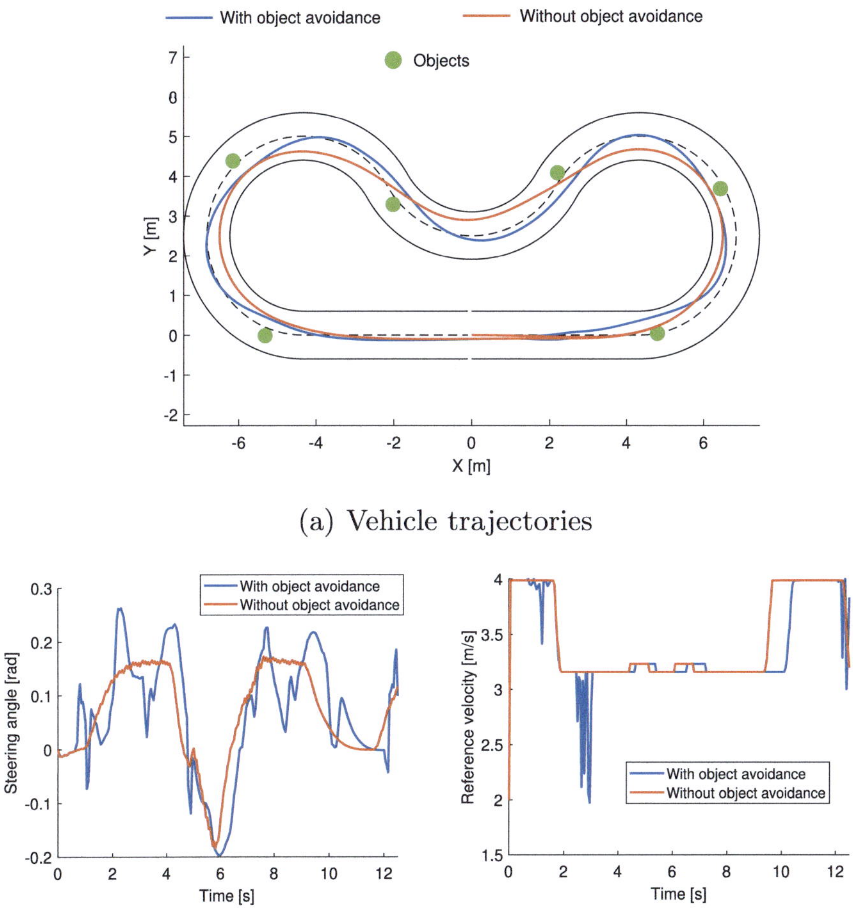

(a) Vehicle trajectories

(b) Reference steering angle

(c) Reference velocity

Fig. 2 Effectiveness of the object avoidance

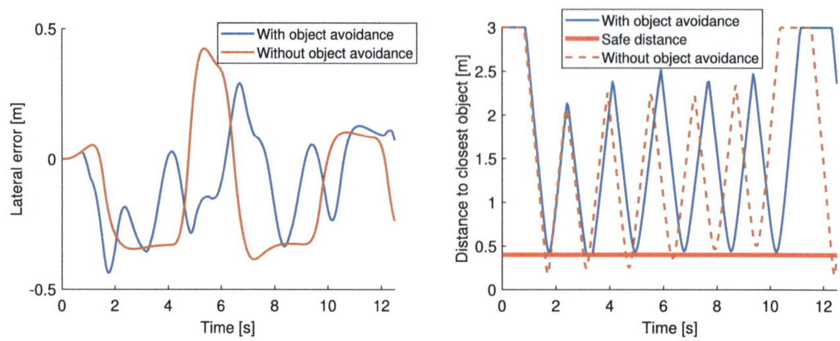

(a) Lateral error of the vehicles (b) Distant to objects in case of the supervised control agent

Fig. 3 Path tracking error and distance to the closest object

maximum of the allowed lateral error is 0.5 m and the safe distance to avoid objects is 0.4 m. A $T = 1$ sec long prediction horizon was chosen with $K = 6$ different control inputs in the ht predictive layer.

The effectiveness of the supervised framework is illustrated in a simulated example, where the vehicles have to complete a full lap on a simple track constructed with line and arc primitives. Along the track, objects are scattered around the centerline. The operation of the supervised control framework is shown with and without the object avoidance predictive layer.

Figure 2 shows the results of the simulations. Both controllers were able to complete a lap on the track, however, the one with object avoidance was able to keep an adequate distance from the objects on the track. Figure 2a–b show the control intervention used by the controllers. The signals are similar, but when objects are detected, the trajectory of the agent is altered to achieve safe motion.

Figure 3 shows the path-tracking error of the controllers and the distance to the closest object along the track. The path-tracking error was kept within the limit in both cases, however, the safe distance to the object required the predictive layer within the supervisor.

The performances concerning lap time are summarized in Table 1. Without object detection, the lap time shortens by approximately 3%, however, in this case, a crash with the objects cannot be avoided. The supervised control system is able to provide almost the same performance with guaranteed safe motion.

Table 1 Lap times in case of the different controllers

Controller	Lap time (s)
With object detection	11.05
Without object detection	11.75

7 Conclusions

The paper has proposed a safe robust framework for high-performance nonlinear control systems of autonomous vehicles. The combination of the reinforcement learning-based control agent and the baseline robust controller was achieved using a supervised control framework. The supervisor contains a model-based predictive layer to be able to predict the vehicle motion and avoid future unsafe situations like leaving the road/track, or crashes with objects, other vehicles, or pedestrians. The effectiveness of the system was illustrated using a simulated example. This capability can be effectively used in different environments with autonomous vehicles, where real traffic is present as well. Possible future work includes the implementation of the proposed algorithm on a real test vehicle and in different simulated traffic scenarios.

Acknowledgements The research was supported by the European Union within the framework of the National Laboratory for Autonomous Systems (RRF-2.3.1-21-2022-00002). The research was supported by the National Research, Development and Innovation Office (NKFIH) through the project 'Design of automated driving systems based on estimation of wheel-road contact features for handling emergency situations' (2019-2.1.11-TÉT-2020-00125).
The work of Attila Lelkó was supported by the ÚNKP-23-3 New National Excellence Program of the Ministry for Innovation and Technology from the source of the National Research, Development and Innovation Fund.

References

1. Sadati N, Berenji H (2016) Coordination of large-scale systems using fuzzy optimal control strategies and neural networks. In: 2016 IEEE international conference on fuzzy systems (FUZZ-IEEE), pp 2035–2042
2. Antonio-Toledo ME, Sanchez EN, Alanis AY (2016) Robust neural decentralized control for a quadrotor UAV. In: International joint conference on neural networks (IJCNN), pp 714–719
3. Wang T, Gao H, Qiu J (2016) A combined adaptive neural network and nonlinear model predictive control for multirate networked industrial process control. IEEE Trans Neural Netw Learn Syst 27(2):416–425
4. Lelkó A, Németh B, Gáspár P (2021) Stability and tracking performance analysis for control systems with feed-forward neural networks. In: European control conference (ECC), pp 1485–1490
5. Ruan W, Huang X, Kwiatkowska M (2018) Reachability analysis of deep neural networks with provable guarantees
6. Huang X, Kwiatkowska M, Wang S, Wu M (2017) Safety verification of deep neural networks

7. Wu M, Wicker M, Ruan W, Huang X, Kwiatkowska M (2020) A game-based approximate verification of deep neural networks with provable guarantees. Theor Comput Sci 807:298–329
8. Németh B, Gáspár P (2021) Guaranteed performances for learning-based control systems using robust control theory. Springer International Publishing, Cham, pp 109–142
9. Németh B, Gáspár P (2021) Ensuring performance requirements for semiactive suspension with nonconventional control systems via robust linear parameter varying framework. Int J Robust Nonlinear Control 31(17):8165–8182
10. Babu VS, Behl M (2020) f1tenth.dev - an Open-source ROS based F1/10 autonomous racing simulator. In: 2020 IEEE 16th international conference on automation science and engineering (CASE), pp 1614–1620
11. Hewing L, Liniger A, Zeilinger MN (2018) Cautious NMPC with gaussian process dynamics for autonomous miniature race cars. In: 2018 European control conference (ECC). IEEE
12. Rosolia U, Borrelli F (2020) Learning how to autonomously race a car: a predictive control approach. IEEE Trans Control Syst Technol 28(6):2713–2719
13. Pacejka HB (2004) Tyre and vehicle dynamics. Elsevier Butterworth-Heinemann, Oxford
14. Schulman J, Wolski F, Dhariwal P, Radford A, Klimov O (2017) Proximal policy optimization algorithms

Image-Based Autonomous Driving and Control System

Vision Based Driving Agent for Race Car Simulation Environments

Gergely Bári and László Palkovics

1 Introduction

Race cars always corner on the tire grip (friction) limit, and vehicle control (driving) on the tire grip limit is important during accident avoidance for self-driving passenger cars. This is a good reason why the development of an algorithm that can drive a real vehicle at the level of a race car driver is an interesting research topic. This paper aims to demonstrate that state-of-the-art RL methods are indeed useful tools for solving this problem, as a fist step, in a simulated environment.

Most articles in this field address the development of classic subsystems according to the problem decomposition, such as trajectory planning or vehicle control. There are less learning-based approaches documented yet; however, their advantages seem to be well documented in the past [3, 9, 10, 14].

In [3] the task of autonomous car racing is solved in a computer game called Gran Turismo Sport. The results show that the obtained controllers not only beat the built-in nonplayer character of the game, but also outperform the fastest known times in a dataset of more than 50,000 human drivers. In the currently available literature, this is the closest approach to that taken in this work. Although the key ideas are similar, the use of visual input in this work introduces significant differences between the two works.

G. Bári (✉) · L. Palkovics
Széchenyi István University, Győr, Hungary
e-mail: gergely.bari@humda.hu

G. L. Conte and O. Sename (eds.), *Proceedings of 12th International Conference on Mechatronics and Control Engineering*, Lecture Notes in Mechanical Engineering,
https://doi.org/10.1007/978-981-96-6452-8_15

2 Modeling Approach

Considering all above, the chosen goal in this work was to use state-of-the-art RL methods, and create an agent, capable of driving in simulation a specific race car, on a specific racetrack with lap times of a professional human driver, while using only visual information (pixels). The task of driving a race car is transformed into a Reinforcement Learning (RL) problem following the general RL approach pictured in Fig. 1.

In this work, the **agent** is a single Neural Network (NN). This NN has a special type, and it is called Convolutional Neural Network (CNN). The (CNN) used here has a structure presented in [5]. The **Action** decided by this agent consists of two real numbers in the range of $[-1..1]$. One of these represents the scale of maximum throttle $(+1)$ to maximum brake, while the other represents the steering wheel angle from maximum left $(+1)$ to maximum right (-1). It is common in such works to limit the agent's control abilities to human level, which usually means the introduction of some kind of rate limit on the actions. In this work, the agent-to-environment interaction happened at 20 Hz, while the human professional race car driver's control frequency is generally considered around 10 Hz. As the main goal of this work was to prove that RL algorithms can indeed create professional level race driving behavior, and not a clear human versus AI competition, this difference was accepted. The used **Environment** is an open-source race car simulation software called TORCS [1]. TORCS, being a detailed car simulation, can provide vehicle dynamic signals and visual 3D representation (pixels) about the vehicle-track environment. In this work these pixels are used as the **State**, while the vehicle dynamic signals are used to form appropriate Reward for training the RL algorithm. This work uses a single NN as agent, so we use a so-called end-to-end method, without further decomposition of the sensor-to-actuator mapping. Although the choice of usual Perception-Planning-Control subsystems seems advantageous in case of road cars, it can lead to some problems in case of racing. In case of road cars, safe trajectories can be planned assuming some worst-case grip values, while in racing, planning a time optimal trajectory would require estimating the grip with very high ($<1\%$) precision, and currently available grip estimation techniques are only in the range of $\sim 10\%$ accuracy [2].

To put the grip sensitivity in perspective, the 1% change in grip level can easily result in 0, 15% lap time difference [4]. In numbers, this means that on a 1 min 40 s lap, dropping average grip level from 1.0 to 0.99 will result in 0.15 s increase in lap

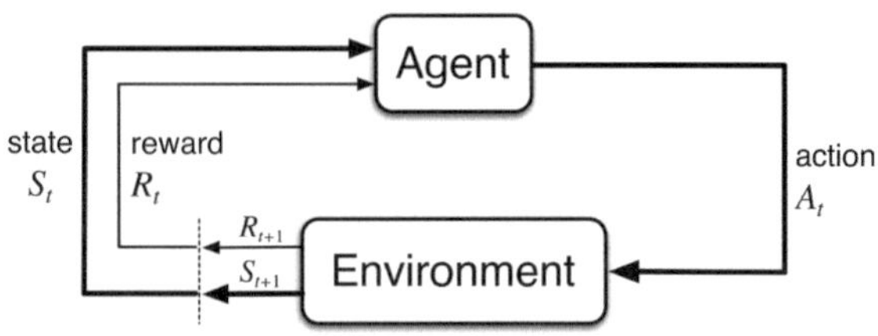

Fig. 1 Scheme of the so called Markov Decision Process, formalizing the Agent–Environment interaction in Reinforcement Learning problems [13]

time, which can easily decide between first and second places. Even if the grip levels (tire model parameters) are completely known, the computational needs to establish the ideal racing line are usually high, or the results are not comparable with those of a human driver [8]. As the grip level cannot be estimated with a good enough precision, the tire model, and hence the whole vehicle model cannot be known too. Therefore, it is reasonable to try the so-called model-free techniques for driver modeling in such setting. This kind of uncertainty and sensitivity transforms the race driving behavior modeling from control design task into a risk management task. The racing driver does not have a clear trajectory in mind during driving. It is continuously trying to find the proper direction and maximum magnitude of the in-plane acceleration of the vehicle, and the trajectory evolves while the boundary of the vehicles' acceleration capabilities is tracked. During this boundary tracking, there is a certain level of risk arising from the fact that the exact grip (the potential of the vehicle to change state) is unknown and the car can slide off the track when crossing its limits. This kind of risk management appears in the evolution of lap times during a qualifying session, too. In these cases, drivers usually have 3–4 attempts to make their quickest lap times, and it is usual that we see laps getting faster and faster from attempt to attempt. There can be technical reasons for this (e.g., track grip evolution, vehicle mass reduction with fuel use, etc.) but the fact that drivers accept higher and higher chances of making mistakes is another key reason for this. Managing this risk is also important to have better and better lap times, and it is done by driver intuition during race car driving. As reports about recent Deep Reinforcement Learning achievements usually report some kind of emergent intuition of these agents [12], so it is straightforward to try modeling this risk management with such deep learning techniques instead of traditional ones.

In Reinforcement Learning **Reward** is a special signal with a numerical value that the agent seeks to maximize over time through its choice of actions. In the present work, the goal is to create an agent with the behavior of minimizing lap time. An important aspect of constructing a reward function is how dense or sparse it is. Using only lap-time as a reward signal is a very sparse reward. It makes learning very unstable and slow, as the feedback about actions comes only after one lap (end of episodes), which can easily be a few thousand steps. To construct a reward that is available at each step (dense), we start by considering how race car drivers learn to drive fast. For race car drivers, the main feedback signal when practicing is a so-called time-difference signal. Usually this value is calculated for every moment and displayed on the driver's dashboard. This signal compares the actual lap to a previous "reference lap" (usually the fastest lap of the driver) and shows the driver how much shorter time it took to reach the actual track position in the actual lap compared to reach the same position in the reference lap. So when drivers experience with a different strategies, (e.g.: different racing lines in a corner), they tend to check if the time difference on their dashboard increased or decreased during the corner. Based on these, chosing the change in the time-difference signal as the basis of the reward function seems a good choice for the purpose of this work. Considering a reference lap that is basically a constant speed movement along the track centerline, and assuming fixed time steps between each Agent-Environment interactions, this reward simplifies to the progress made in each time step. It worth noting that this thinking, in principle, leads to the same reward used in [3, 14].

$$r_t^{t_{diff}} = s_t^{cl} - s_{t-1}^{cl} \tag{1}$$

where s_t^{cl} is the distance traveled along the track centerline in timestep t.

To make training more robust, additional components were introduced into the final reward value. One component was defined depending on the reason why a learning episode was terminated (r_t^{ter}). In this, a constant value was added to the reward if the agent finished the lap and penalties were introduced for the reasons: leaving the track, "turning back" on the track, if the car damaged, progressed "backwards" on the track, or the agent made too small progress in a given time window. In addition, a reward component was added to punish the agent using "too high" action values (r_t^{act}). As the neural network output can be any value, it was found during training that adding a reward component that punishes non-feasible, out-of-bound actions helped stability of training.

$$r_t^{act} = (\frac{|a_t|}{p^{sc}} - p^{bnd} + 1)^2 \tag{2}$$

The final reward function is then:

$$r_t = r_t^{t_{diff}} + r_t^{ter} - r_t^{act} \tag{3}$$

Specific values for these parameters are summarized in Table 1.

For training the agent the Proximal Policy Optimisation (PPO) algorithm was used [11]. PPO is a well-known, widely used on-policy, model-free RL algorithm. In recent years, it has proved its robustness in various use cases [6, 7]. For gathering experiences for training multiple (32) TORCS instances were run in parallel. Trainings with various hyper parameters were performed on DELL R730 E5-2670 v3, computers. Trainings were stopped after a maximum of 10 days, which corresponds approximately one billion training steps in this case. The performance of the agent was monitored during the training and in every 10000 step a test episode was performed. In this case the mean value of the agent CNN stochastic output (actions) was used, while during learning actions were sampled from these stochastic variables. Training parameters for the results presented in this work are summarized in Table 2.

3 Simulation Results

To efficiently show important aspects of the results, a specific figure layout is used in this section, so Figs. 2, 3, and 5 share the same structure. The top graph shows how far the agent can drive on the track as a function of the learning steps, while

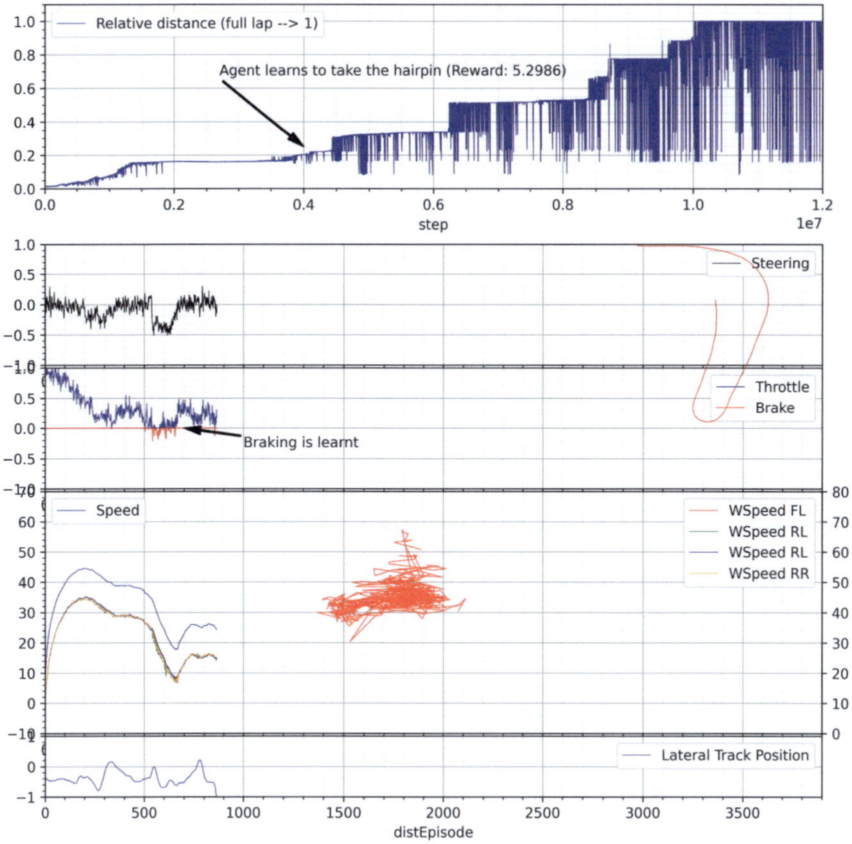

Fig. 2 An episode showing that the Agent learns breaking and takes the hairpin, but fails to turn left in the following (first left) corner

the bottom graphs show signals as a function of the traveled distance. This second part shows on the top the "driver inputs", eg.: the steering and the throttle/brake signals (+1 means full throttle, −1 means full brake pedal application), then the graph below shows vehicle speed with wheel speeds in [m/s] (y axis is in offset), and the last graph on the bottom shows track position, where +1 refers to the left, −1 the right side limit of the racetrack. There is also a red "patch" in the middle of the speed trace. This patch is created by plotting the longitudinal and lateral acceleration of the vehicle. In racing terms, this is the so-called "GG diagram". Having a nice "round" or "fat" GG diagram generally shows that the car is driven close to its limits, while a GG diagram with a "cross-like" or "thin" shape usually refers to a more slow, road-car-like driving behavior. To better understand the location of the track, the plots also show the traveled trajectory of the vehicle, as an overlayed single line over the steering, throttle/brake, and speed diagrams.

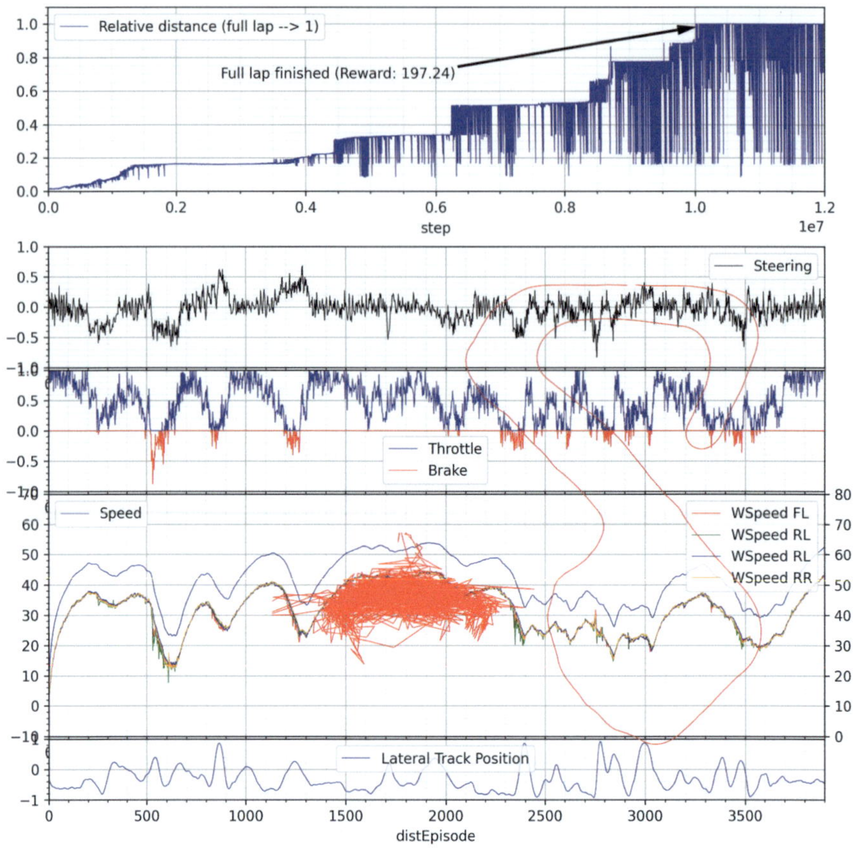

Fig. 3 An episode, showing Agent learned to drive the full lap for the fist time

Several learning runs were performed in this work. Trials with various reward functions or hyperparameters resulted in more than a hundred runs, and most of these runs lasted hundreds of millions of learning iterations. However, in all of them, the agent learned relatively quickly in ~10 million steps how to complete a full lap, and during this phase the learning showed the same pattern. The interesting aspect of this pattern is described in the following.

In Figure 2 it can be seen that at the beginning of the training process the unlearned agent shows random behavior and shortly after starting the episode it leaves the track. This is represented in the top graph, as 0 learning steps corresponds to 0 distance. As learning steps grow, the distance traveled also increases. This shows how the agent learns to drive straight. The first "plateau" in the traveled distance starts at ~ 0.5 million steps, slightly less than 20% of the total lap distance (the relative distance is ~0.2). Here, the agent learns to steer according to the first (almost straight, full throttle, relatively "easy") corner, while at ~4.5 million steps and slightly more than or 20% of the total lap distance, the braking behavior emerges. Here the agent reaches a hairpin, that is a small radius, long-arc corner, which can only be performed with small speed. Therefore, the agent must learn that at some point it needs to apply brake instead of throttle to keep the car on track. Note that this plateau is quite long. It took ~4 million steps (~40%) to learn breaking and take this single corner.

Figure 3 shows an episode created at ~10 million learning steps, where the agent learned to go around the full lap for the first time. Keeping focus on the top graph, the next "step change" in the traveled distance comes ~6.2 million learning steps. Here, the agent learns to turn left and progresses to the next sequence of corners. Up until this point, the track had only right corners (first corner and the hairpin), so taking these two left-handers required learning a new behavior. Note that until ~8.3 million steps, the agent does not learn to process further. Here, the agent has already spent 80% of its "full lap learning time" (~8 of ~10 million steps), but can only reach ~ 50% of the lap yet. This pattern of learning makes perfect sense. After long learning to accelerate, brake, turn right and left, the agent will learn faster the remaining corners, almost a quarter of the steps (~2 million steps). The next notable milestone is how the agent learned a fast right-hand sequence of the track (~8.5 million steps). These corners are very fast with lots of bumps. This is a very tricky part of the track, as the fast speed combined with the bumpiness makes the car unstable. So although the agent has already learned to turn right in the hairpin before, these right corners require totally different behavior. It is also noteworthy that the agent basically learned this sequence at once; after it learned to take the first of these kinds of turns, it passed all the others too with almost no learning.

The agent also needs to take some steps to learn the last left-hand turn after the quick right corner sequence (~9.5 million steps). Although this corner is a left-hander and seems the same type as the previous two left-handers, this corner is still a "new" type thanks to the preceding quick right turns that affect the ideal line for this corner. The short straight before this corner requires to start braking approximately in the middle of the track instead of the usual choice of the outside edge. Consequently, even though this is not the first left corner to be learned, the approach is still novel and, as a result, it takes approximately one million steps to learn this turn, which is a considerable amount of time at this relatively late stage of the learning process.

In the final phase of this learning, at ~10 million learning steps, the agent can drive around the lap, however, the plots in Fig. 3 show that it is not yet driving at the grip limit. One clear sign of this is the pattern in the throttle/brake input curve. Racing-like driving usually means constant full throttle application in straights, with sudden change to high brake pedal application before corners that is not seen here yet. Furthermore, the GG diagram (red patch) shows low braking dynamics because the shape is not round.

After learning to drive around the track in ~10 million steps, it took 100 times more steps (~1000 million steps) for the agent to evolve to the professional human level. Figure 4 shows this disproportion nicely.

Figure 5 shows a driving behavior that is very similar to that of a professional race car driver. The patterns on the graphs are almost identical to those recorded from real race car drivers, which is further corroborated by interviews with experts (race engineers and professional drivers). The only trace that is noticeably different from

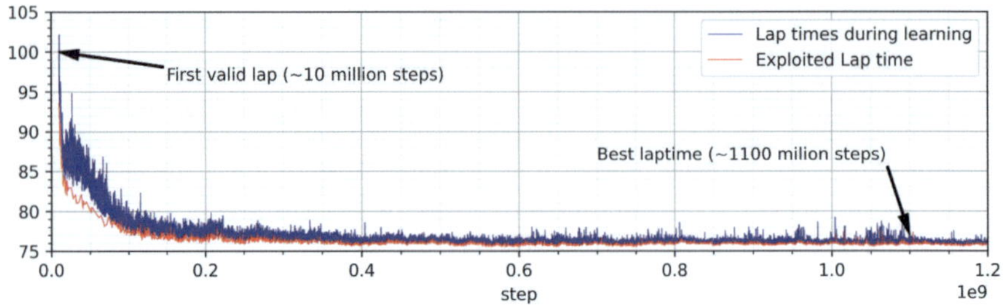

Fig. 4 Lap time evolution during learning

that of human drivers is the steering wheel angle trace. This is to be expected, as the agent in this work was not subject to any rate limit, and the power needed to turn the steering wheel was not modeled. Considering throttle/brake signal, the pattern here is indistinguishable from a human race driver based on human experts feedback. In a straight line there is full throttle application (flat sequences at maximum (+1)). When the agent reaches the braking point before a given corner, this maximum throttle turns into maximum brake signal (maximum as maximum possible brake without locking the wheels). Note that this is also a common pattern in classical time-optimal control problems. The GG diagram in Fig. 5 is also a key indicator of grip limit maneuvering, as it has a nice round shape in this case.

Note one really interesting aspect of this behavior. The agent learns how to brake on the tire grip limit without blocking the tires, although only visual information is available, without direct wheel speed signals. This is clear by checking the wheel speed traces in Fig. 5. There are noise-like spikes during braking on the wheel speed traces. This indicates that the wheels are in the locking limit, but none of them actually stops rotating (drops to 0 speed). This pattern in wheel speed then shows up in the brake pedal usage too. Where the wheels show a locking pattern (some wheels start to slow down more than others), the agent decreases the braking effort.

Learning this antilock-like behavior without direct wheel speed information is not straightforward, as wheel speed is generally considered essential for Vehicle Dynamic Controls (VDC) functionalities like Anti Block System (ABS). One possible reason for this is that the agent has not direct (wheel speed) but indirect information about wheel locking. It is an essential property of tires, that when they are locked they lose the capacity to create lateral forces, thus control through steering is lost in such situation. Therefore, although the agent does not know the rpm of the wheels, it can recognize the loss of steering control when the wheels are blocked. As the expected episode reward will be less in these cases, this feeds back into learning to avoid these situations by modulating brake pressure to avoid wheel lockups.

An other interesting result is that the agent learns to select an optimal racing line through the corners, as seen in the bottom graph of Fig. 6. (This is known as "trajectory planning" in classical self-driving terminology). It can be seen that before the agent starts braking, it approaches the outside edge of the track. Then it approaches the inside edge, approximately when it reaches the minimum speed (in racing terms:

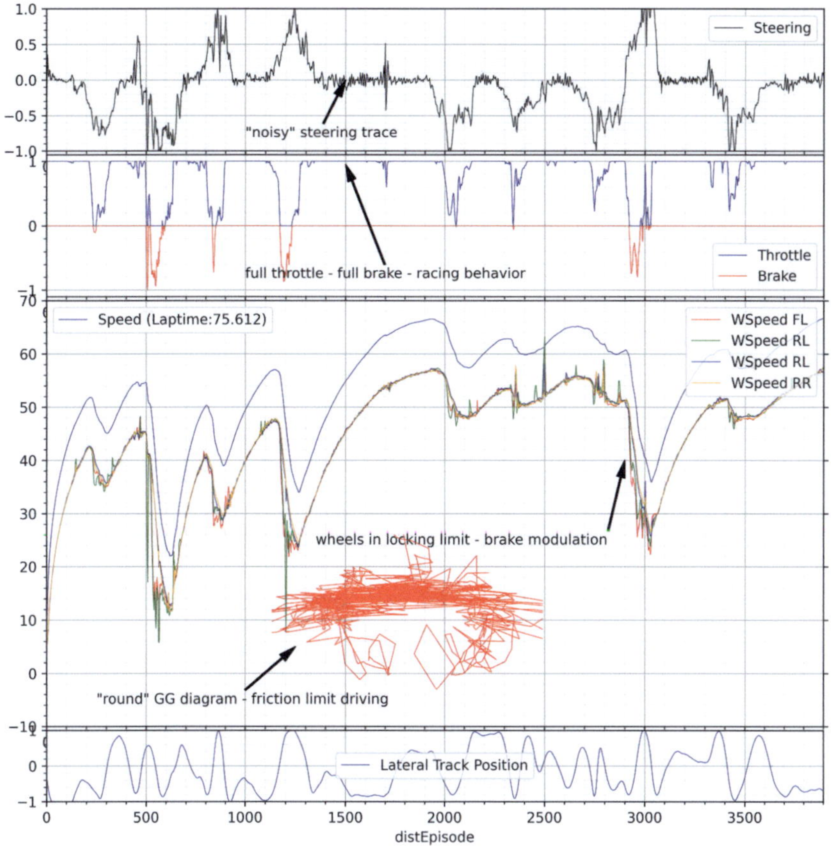

Fig. 5 A plot showing human pro-like driving behavior

it reaches the apex), and lastly it "falls" to the outside edge of the track again when it applies maximum throttle.

4 Conclusion

This work demonstrates the effectiveness of deep reinforcement learning algorithms in learning self-driving behaviors without prior knowledge or experience. Results show that self-driving on the tire grip limit can be formed as a reinforcement learning problem and that Proximal Policy Optimization is suitable to solve this problem using visual state information.

It would be interesting to investigate the effects of allowing an AI agent to control all four wheels independently, with the ability to adjust the torque and steering angle

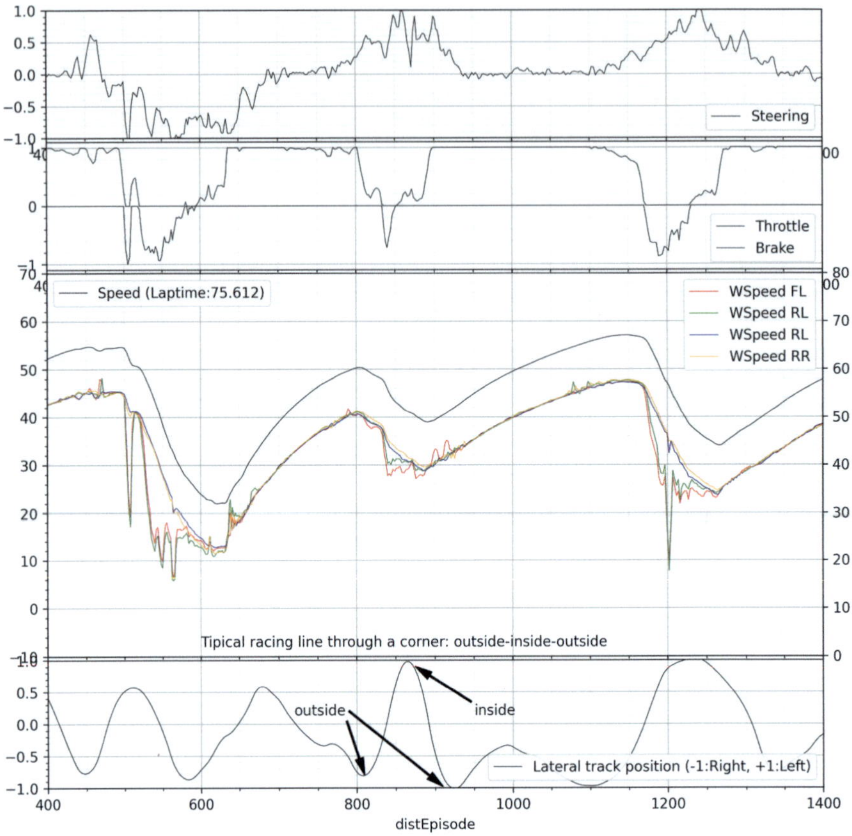

Fig. 6 Plots showing how the agent chooses racing line in a corner

of each wheel, as well as to compare the agent's performance with a human driver in both simulated and real-world scenarios. Additionally, it would be beneficial to research the behavior of the agent if the controlled car has Anti-lock or Traction control systems.

Acknowledgements The research was supported by the European Union within the framework of the National Laboratory for Autonomous Systems (RRF-2.3.1-21-2022-00002).

5 Appendix

See Tables 1 and 2

Table 1 Parameters for the reward function in Eq. (3)

Parameter	Value
Reference Speed (v^{ref})	20
Action scaling (p^{sc})	15
Bound for scaled action (p^{bnd})	1.2
Termination reward components (r_t^{ter}):	
Reward for reaching the finish (`dist_episode` > `3900`)	+100
Punishment for leaving the track (|`track_pos`| > `1.2`)	−10
Punishment for turning back (`angle` < `0`)	−10
Punishment for damage the car (`damage` > `0`)	−10
Punishment for progress backwards (`damage` > `0`)	10
Punishment for low progress (`timestep` > `500` AND `episode_reward` < `0`)	−10

Table 2 Training hyperparameters for Stable Baselines PPO, and the TORCS simulator environment

Parameter	Value
Learning rate—with decay	$[1 : 2.5 \cdot 10^{-4}, 0 : 0.5 \cdot 10^{-4}]$
Maximum training steps	$1.5 \cdot 10^9$
Discount factor	0.995
Entropy coefficient	0.01
Value function coefficient	0.5
Policy clip range	0.2
Value function clip range	0.2
Environment instances	24
Batch size	512
Agent-Environment interaction timestep (ts)	0.05 s
TORCS parameters:	
Used track name	brondehach
Used car model name	155-DTM
`ASR_ON`	False
`ASR_ON`	False
Display mode constants:	
`RM_DISP_MODE_PYTORCS_FIXFPS`	8
`RM_DISP_MODE_PYTORCS_DISPLAY`	16
`RM_DISP_MODE_PYTORCS_CAPTURE`	32
Observation resolution	$84 \times 84 \times 84$, grayscale
Timestep in physics simulation	0.002 s

References

1. (2020) TORCS - The Open Racing Car Simulator. https://sourceforge.net/projects/torcs/
2. Acosta M, Kanarachos S (2018) Tire lateral force estimation and grip potential identification using neural networks, extended kalman filter, and recursive least squares. Neural Comput Appl 30(11):3445–3465. Springer
3. Fuchs F, Song Y, Kaufmann E, Scaramuzza D, Duerr P (2020) Super-human performance in gran turismo sport using deep reinforcement learning. arXiv:2008.07971 [cs]
4. Kelly DP, Sharp RS (2010) Time-optimal control of the race car: a numerical method to emulate the ideal driver. Vehicle Syst Dyn 48(12):1461–1474. https://doi.org/10.1080/00423110903514236. Taylor& Francis
5. Mnih V, Kavukcuoglu K, Silver D, Graves A, Antonoglou I, Wierstra D, Riedmiller M (2013) Playing atari with deep reinforcement learning. arXiv:1312.5602
6. OpenAI, Akkaya I, Andrychowicz M, Chociej M, Litwin M, McGrew B, Petron A, Paino A, Plappert M, Powell G, Ribas R, Schneider J, Tezak N, Tworek J, Welinder P, Weng L, Yuan Q, Zaremba W, Zhang L (2019) Solving Rubik's Cube with a Robot Hand. arXiv:1910.07113 [cs, stat]
7. OpenAI, Berner C, Brockman G, Chan B, Cheung V, Dębiak P, Dennison C, Farhi D, Fischer Q, Hashme S, Hesse C, Józefowicz R, Gray S, Olsson C, Pachocki J, Petrov M, Pinto HPdO, Raiman J, Salimans T, Schlatter J, Schneider J, Sidor S, Sutskever I, Tang J, Wolski F, Zhang S (2019) Dota 2 with large scale deep reinforcement learning. arXiv:1912.06680 [cs, stat] http://arxiv.org/abs/1912.06680
8. Perantoni G, Limebeer DJ (2014) Optimal control for a formula one car with variable parameters. Veh Syst Dyn 52(5):653–678. Taylor & Francis
9. Remonda A, Krebs S, Veas E, Luzhnica G, Kern R (2021) Formula RL: deep reinforcement learning for autonomous racing using telemetry data. arXiv:2104.11106 [cs]
10. Remonda A, Veas E, Luzhnica G (2021) Acting upon Imagination: when to trust imagined trajectories in model based reinforcement learning. arXiv:2105.05716 [cs]
11. Schulman J, Wolski F, Dhariwal P, Radford A, Klimov O (2017) Proximal policy optimization algorithms. https://doi.org/10.48550/arXiv.1707.06347 [cs]
12. Silver D, Huang A, Maddison CJ, Guez A, Sifre L, van den Driessche G, Schrittwieser J, Antonoglou I, Panneershelvam V, Lanctot M, Dieleman S, Grewe D, Nham J, Kalchbrenner N, Sutskever I, Lillicrap T, Leach M, Kavukcuoglu K, Graepel T, Hassabis D (2016) Mastering the game of Go with deep neural networks and tree search. Nature 529(7587):484–489. https://doi.org/10.1038/nature16961, number: 7587. Nature Publishing Group
13. Sutton RS, Barto AG (2018) Reinforcement learning: an introduction. MIT Press
14. Wurman PR, Barrett S, Kawamoto K, MacGlashan J, Subramanian K, Walsh TJ, Capobianco R, Devlic A, Eckert F, Fuchs F, Gilpin L, Khandelwal P, Kompella V, Lin H, MacAlpine P, Oller D, Seno T, Sherstan C, Thomure MD, Aghabozorgi H, Barrett L, Douglas R, Whitehead D, Dürr P, Stone P, Spranger M, Kitano H (2022) Outracing champion Gran Turismo drivers with deep reinforcement learning. Nature 602(7896):223–228. https://doi.org/10.1038/s41586-021-04357-7

Design of a Physiological and Vehicle-Oriented Data Acquisition Platform for Autonomous Vehicle Users

Rebeca Mellado-Reyes, Jesús A. Mandujano-Granillo, Alejandro Arceo, Luis C. Félix-Herrán, Mauricio A. Ramírez-Moreno, Juan C. Tudón-Martínez, and Jorge De-J. Lozoya-Santos

1 Introduction

Nowadays, the production and creation of autonomous vehicles has been a rising tendency that does not seem to be stopping any time soon. However, the implementation of these type of vehicles is still in developing stages, as security refinements and reinforcements need to be made to ensure the safety of the passenger or driver. Not forgetting that many users might feel uneasy as they do not trust autonomous driving vehicles. Therefore, new ways to ensure user well-being, comfort, and security need to be implemented, and the analysis of biometric and physiological data is an approach that can be taken to resolve this matter.

As a result of this presented problem, a platform was designed with the objective of being an integration system where physiological and vehicle data sensors acquire and record data simultaneously of a user during a route made by an autonomous vehicle. In this way, a medium and integration platform can be offered to acquire data for further analysis to measure quantitatively, through biomedical sensors, the user-vehicle experience.

Other related works have outlined the importance of having biometric and behavioral data recollected from users in automotive systems, but these are not as personalized as they should be in regard to providing data that could potentially save their lives, for example [1]. On the other hand, outlining the feasibility of these type of

R. Mellado-Reyes · J. A. Mandujano-Granillo · A. Arceo · L. C. Félix-Herrán · M. A. Ramírez-Moreno (✉) · J. C. Tudón-Martínez · J. D.-J. Lozoya-Santos
Mechatronics Department, School of Engineering and Sciences, Tecnologico de Monterrey, Monterrey, Mexico
e-mail: mauricio.ramirezm@tec.mx

R. Mellado-Reyes
e-mail: a01198024@tec.mx

G. L. Conte and O. Sename (eds.), *Proceedings of 12th International Conference on Mechatronics and Control Engineering*, Lecture Notes in Mechanical Engineering,
https://doi.org/10.1007/978-981-96-6452-8_16

platforms, data from monitoring systems with remote sensors has also been implemented to make the relationship between limited physiological data and vehicle performance [2]. The results showed the correlation between driver and vehicle variables such as CO_2 consumption, driving styles, and physiological data. In contrast to these alternative platforms, the one designed adopts a personalized approach, offering a flexible and universal user-testing medium that recurrently stores data for subsequent quantitative analysis of the user's experience.

The outline of this work is as follows: Section 2 presents the methodology used in the proposed system, where the materials, experimental design, and data processing are described. Section 3 is devoted to presenting the preliminary results of this current work. Finally, Sects. 4 and 5 conclude the work and propose future work, respectively.

2 Methodology

The methodology proposed in this paper represents the acquisition of the data of a user and vehicle during a route. This includes biometric wearable and noninvasive devices used to acquire physiological data, while vehicle-oriented data is obtained from integrated and programmable sensors in the vehicle that output variables related to the driving cycle. In addition, it integrates the use of hardware and software components to conclude with raw data files that can be analyzed in further work.

2.1 Vehicle Route Specifications

The vehicle employed is the *NEV Polaris Gem e6*, a programmable low-speed vehicle (up to 40 km/h) manufactured by *GEM* in Anaheim, California. Its main components consist of a drive-by-wire system that permits the actuators to be controlled via electronic signals programmed to its main computer, two LiDAR sensors that detect their surroundings through the reflection of laser beams, a GPS/GNSS and IMU positioning system that utilizes satellite location services to identify the vehicle's position and derive speed variables, and two external cameras. The vehicle also has an integrated state machine that provides the vehicle's own variables and commands.

The design of the autonomous driving system (ADS), meaning the route and parameters defined for this study, is delimited by the Operational Design Domain (ODD) standard, defined by the Society of Automobile Engineers (SAE). In this case, a level 3 ADS, according to the SAE's J3016 standard [3], was implemented, as the driver does not engage with the autonomous driving features unless required. The ODD employed includes a vehicle that is automated to drive at 7.2 km/h in laps within a small internal parking space in the university *Tecnologico de Monterrey* (see Fig. 1).

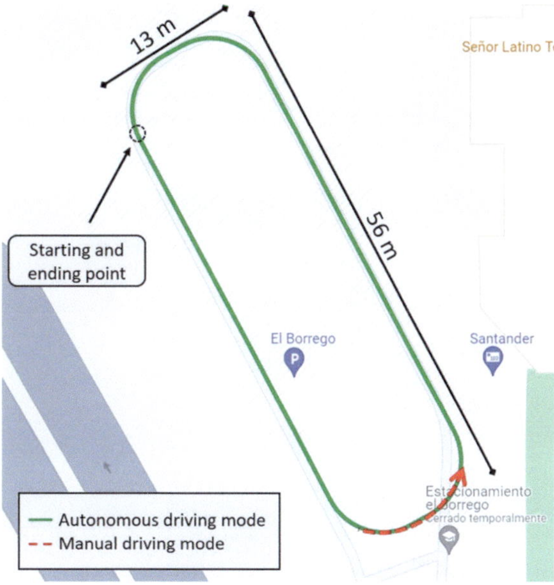

Fig. 1 Route followed by the vehicle, where the red, dashed section represents a section where manual driving mode had to be enhanced due to programming limitations: the vehicle sensors did not recognize the surrounding vehicles when making a close turn

The figure was taken from Google Maps, from where the distance was also measured. Additionally, parameters such as the scene, environmental conditions, and dynamic elements for the ADS are considered as follows:

- Internal areas of the city and a university with capacity for vehicle traffic (not usually) and non-flat, textured (not concrete) streets and buoys, possibly with structural damage (i.e., potholes).
- Clear weather (but not in extreme heat) or cloudy (no rain or thunder); natural lighting.
- No presence of automobile traffic, the experimentation will wait until the car exits the parking lot; occasional pedestrians and animals (i.e., peacocks) might cross unsuspectingly.

2.2 *Experimental Setup*

The system includes *GEM*'s autonomous vehicle, NEV; the Empatica E4 bracelet, manufactured by *Empatica* in Milan, Italy; and the ENOphones, manufactured by *ENO*. The prior records Blood Volume Pulse (BVP) through photoplethysmography (PPG), Inter-beat Intervals (IBI), electrodermic activity (EDA), body temperature at, and acceleration data at. BVP is collected using a PPG sensor that measures changes in blood volume within arteries and capillaries by emitting infrared light through tissues. The amount of light detected by the PPG sensor's photodetector is directly linked to the blood volume in the tissue [4].

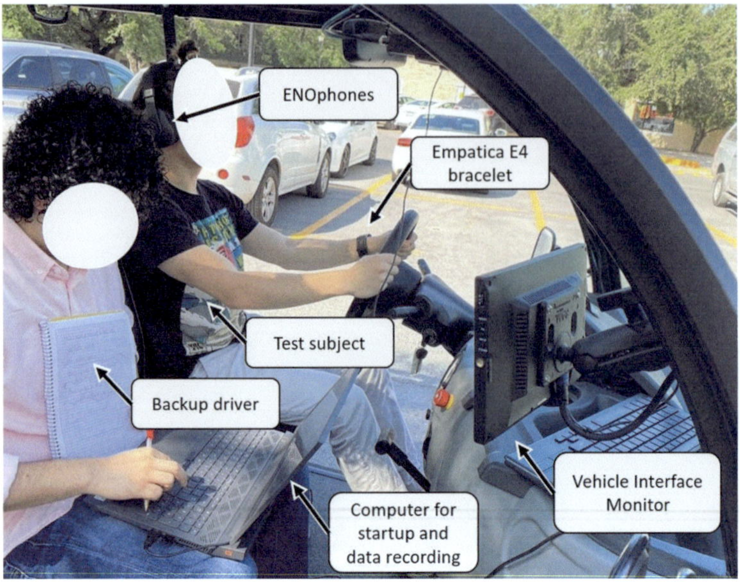

Fig. 2 Experimental setup consisting of a test subject wearing the data collection devices, in this case driving, with a backup driver that manages the vehicle's interface, which can be visualized with the monitor

EDA refers to alteration in the skin's electrical properties resulting from sweat gland activity or perspiration, which is quantified by measuring skin conductance rate (SCR). In analogous research, EDA has been employed to evaluate attention and arousal levels, particularly in assessing driver attention [5].

On the other hand, the Enophone is a four-electrode device that records electroencephalography (EEG) data. These signals can be classified into frequency bands: delta (0.5–4 Hz), related to deep sleep; theta (4–8 Hz), related to a low-conscience, meditative state; alpha (8–13 Hz), related to mental and physical relaxation states; beta (13–32 Hz), related to conscious and alert states; and gamma (32–100 Hz), related to problem-solving and learning [6].

The experimental setup also includes an external computer to enable and control the autonomous mode, and collect the physiological data being obtained by the wearable devices (see Fig. 2). There is a human driver from whom physiological data is being recorded, and a backup driver, however, they can be switched depending on the test being made.

2.3 *Experimental Design*

Before the start of the experiment, the user was placed the biometric devices: making sure the Enophone electrodes have direct contact with the skin by separating strands of hair, and cleaning with rubbing alcohol in the part of the wrist where the Empatica electrodes will be placed. The latter must feel tight but not uncomfortable, otherwise the recording will be compromised. Secondly, a neutral data recording was taken place to ensure the data was correctly saved. Thirdly, the user was asked to take a seat inside the vehicle, and the start of the test will be taken place. The general experimental protocol included a 30-s prior calibration test followed by 2–3 laps around the parking lot, and ending with a 10-s backup period to avoid any data loss, lasting in total 150 s at most. These include two different tests where the test subject was the pilot (in regular and under stress conditions), and an autonomous driving test.

2.4 Data Collection

From each test, the physiological data collected with their own sampling rates are as follows: BVP at 64 Hz, IBI at 2 Hz, EDA at 4 Hz, body temperature at 4 Hz, acceleration at 32 Hz, and EEG at 250 Hz. These were collected synchronously with a Python code that established a connection with the Empatica server through a Wi-Fi connection, and Bluetooth connection with the ENOphones. These codes can be found in Appendix A. The codes resulted in different comma-separated value (CSV) files for each variable with its corresponding timestamp.

On the other hand, the variables obtained from the NEV depended on the vehicle input data, and will be returned as feedback data messages during the autonomous or manual driving condition. In this case, the parameters that result from the data input are steering wheel angle, throttle pedal, brake pedal, and vehicle longitudinal velocities with the vehicle internal sensors and Global Navigation Satellite System (GNNS) velocity estimations. These data messages are recorded as "*rosbag*" files, and processed as time series objects to be able to be plotted and compared.

3 Preliminary Results

Two types of data were acquired from the platform: physiological and vehicle oriented. Data files were obtained for each variable, and these were plotted against time. EEG data of a 115-second test can be visualized in Fig. 3, where the signals collected by the four electrodes are plotted. Commonly, the voltage of EEG data ranges from 10 to 100 μV, which is where the graph is most condensed prior to further processing. Therefore, this demonstrates the proper data recollection of EEG data. On the other hand, Fig. 4 displays raw biometric data collected from that same test, as the EEG data. In this graph, BVP, IBI, and EDA data represents the raw physiological data that can be processed henceforth and analyzed to obtain variables such as heart rate. Firstly, the BVP and IBI subplots demonstrate the cardiac tendencies obtained. The latter shows the data obtained from the PPG, and represents the time

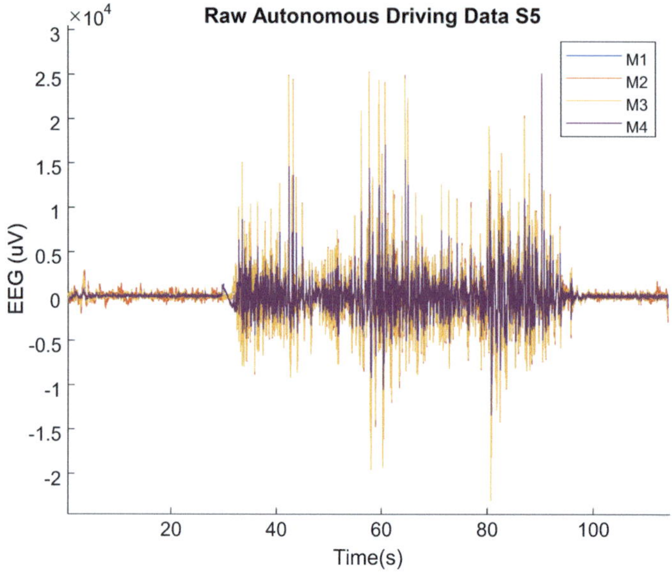

Fig. 3 EEG raw data obtained from the ENOphones during a 115-second autonomous driving test in the 4 electrodes. Each color shows the data of these electrodes

taken between beating intervals. All these variables are in normal ranges. Consequently, the biometric data obtained can be analyzed parallel to that of the driving cycle. Biometric data processing exceeds the limitations of this platform, as the only objective is to recollect in-range data.

On the other hand, the data was successfully obtained from the vehicle, as seen in Fig. 5. This graph shows the comparison between the given, the desired, and the measured value. There is an interruption in all graphs at second 47, where the autonomous mode had to be paused when turning in a parking lot that had other vehicles and the code was limited to properly making the turn without causing damage to other vehicles. What's important about this graph is the representation of the values when making a turn (wheel angle), when braking (pedal break), when accelerating (throttle pedal), and the speed of the vehicle with two different acquisition methods.

To further detail the characteristics of the driving cycle shown in Fig. 5, several phases are shown. The initial calibration phase is visualized from the lack of vehicle movement. Then at the 30-second mark, the autonomous driving mode is activated, in which the vehicle speed is 2 m/s which causes the throttle pedal to be activated at around 60%, to then be reduced to 25% when the velocity is achieved, during this period of initializing the route in autonomous mode. Near the 50-second mark, the vehicle's autonomous driving mode is disengaged, and manual control is used during the route to properly avoid a tight curve, it is observed that the desired steering angles and vehicle pedals go to 0 during this manual driving mode, although the desired speed signal doesn't share this behavior due to delays in the reception of driving switch modes, but this doesn't affect the performance of the vehicle. After the curve

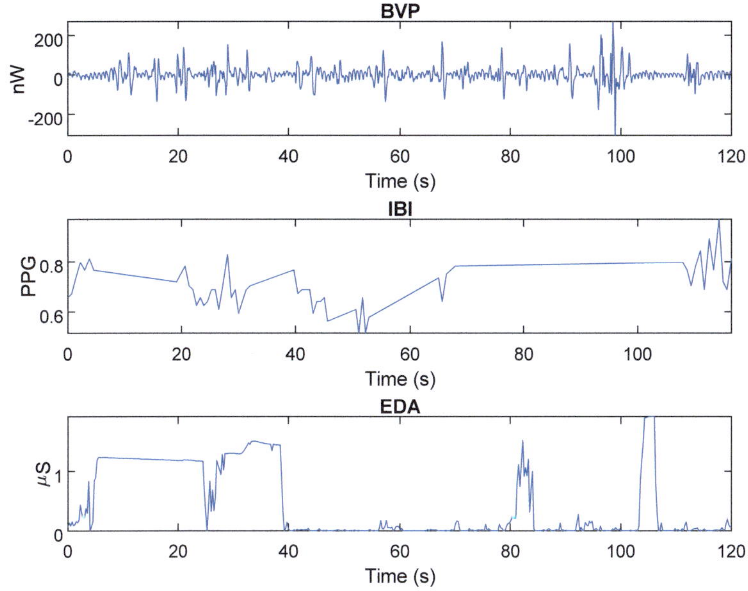

Fig. 4 Raw BVP, IBI, and EDA data obtained from the Empatica bracelet during a 60-second driving test

is passed via manual driving, at the 60-second mark, the autonomous mode is then reactivated up until the end of the route. After the route is completed, the vehicle is stopped, although without the use of the brake pedals, and the user biometric was still recorded during an additional period.

4 Conclusions

The results from this platform include the integration of two biometric devices that in synchrony record data with different sensors at different sampling frequencies. These go together with the resulted data obtained from the autonomous vehicle, which were also within the time window specified for the tests.

Considered the interest and objective defined for this platform, the results collected were enough and qualify for subsequent analyses. However, as the processing and analysis of biometric or vehicle-oriented data is not made within this platform, noise cannot be quantified or eliminated, compromising the visualization of the data from the biometric figures. Another limitation is the amount of variables recollected and the route taken by the vehicle: only those mentioned were obtained and only one vehicle path was followed. This was established in this manner to preserve uniformity and coherency with the interests of this platform: obtaining and storing data from users in autonomous vehicles.

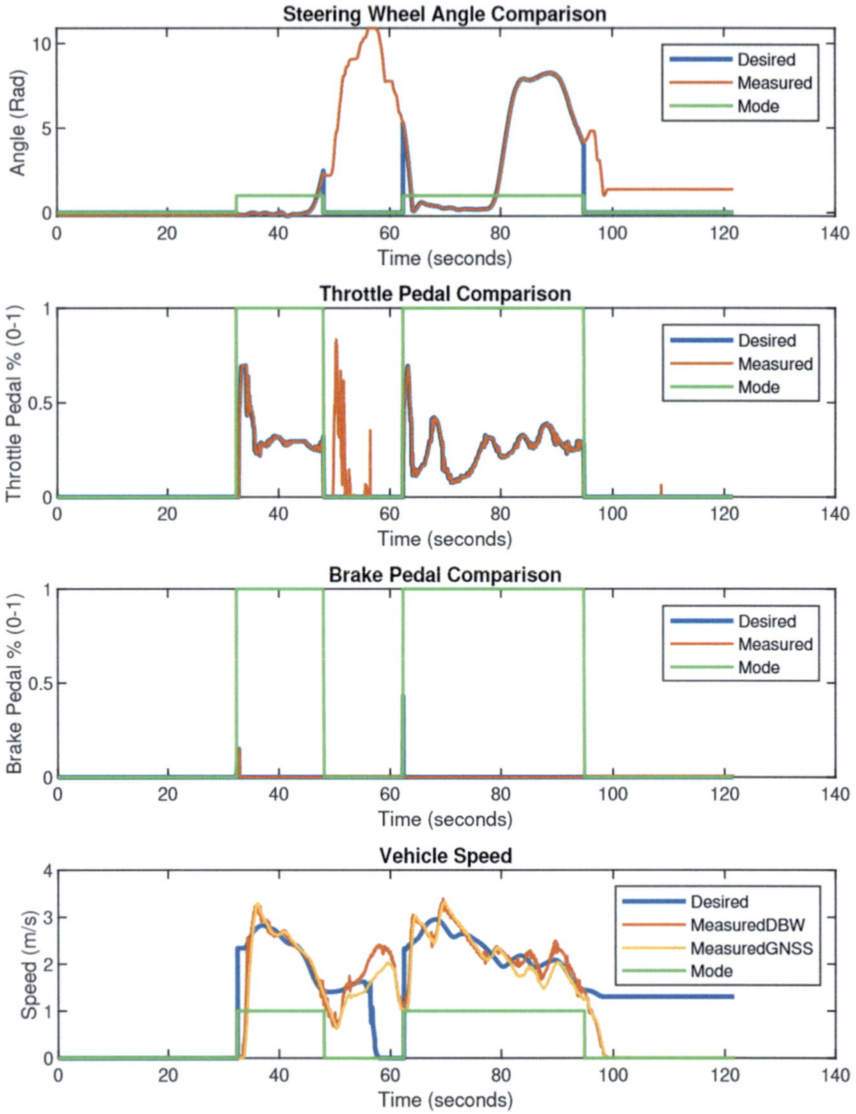

Fig. 5 Vehicle data obtained from the *ros* files, where higher values are those of the actuators being used, while values closer to zero indicate no use of these

Altogether, the results show that the platform can be used for further physiological data analysis where different scenarios are presented.

5 Future Work

Finally, we highlight that the methodology employed in this contribution has the potential for expansion in at least two directions.

- Integrating a harmonious blend of hardware and software components to produce raw data files, facilitating future analysis in subsequent endeavors.
- Finalizing the development of the Biometric Cabin to mitigate car accidents arising from the mentioned emotional and physiological states.

Acknowledgment The authors thank the important technical support made by Enrique Bardawill, Francisco Curiel, Alejandro Rodriguez, and David Gutiérrez, who are Tecnológico de Monterrey's graduate students.

References

1. Lozoya-Santos JdJ, Sepúlveda-Arróniz V, Tudón-Martínez JC, Mendoza RAR (2019) Survey on biometry for cognitive automotive systems. Cogn Syst Res 55:175–191. https://doi.org/10.1016/j.cogsys.2019.01.007
2. Campos-Ferreira AE, Lozoya-Santos, JdJ, Tudón-Martínez JC, Mendoza RAR, Vargas-Martínez A, Morales-Menendez R, Lozano D (2023) Vehicle and driver monitoring system using on-board and remote sensors. Sensors 23:814. https://doi.org/10.3390/s23020814
3. SAE International (2021) Taxonomy and definitions for terms related to driving automation systems for on-road motor vehicles, April 30. https://saemobilus.sae.org/content/J3016202104
4. Peper E, I-Mei L, Harvey R (2007) Is there more to blood volume pulse than heart rate variability, respiratory sinus arrhythmia, and cardiorespiratory synchrony? Biofeedback 35(2):54–61. https://www.researchgate.net/publication/
5. 259560204 Is there more to blood volume pulse than heart rate variability respiratory sinus arrhythmia and cardiorespiratory synchrony
6. Aminosharie T, Affanni A, Rinaldo R, Zontone P (2022) Driver attention assessment using physiological measures from EEG, ECG and EDA signals. Sensors 23(4). https://doi.org/10.3390/s23042039
7. von Stein A, Sarnthein J (2000) Different frequencies for different scales of cortical integration: from local gamma to long range alpha/theta synchronization. Int J Psychophysiol 38(3):301–313. https://doi.org/10.1016/S0167-8760(00)00172-0

Vision System Based on Artificial Intelligence for Service Robots

Carlos E. Mendoza-Ramirez, Juan A. Gonzalez-Aguirre, Juan C. Tudon-Martinez, Luis C. Felix-Herrán, Mauricio A. Ramírez-Moreno, Jorge De-J. Lozoya-Santos, and Olivier Sename

1 Introduction

Vision systems have had a presence within the industry and technological areas for decades, evolving exponentially with the emergence of new technologies [1], such as Artificial Intelligence methods including Machine Learning and Deep Learning techniques [2]. These types of systems began to be more efficient with the development of computer engineering in the 1980s, a time in which faster and more sophisticated processors also emerged [1]. These types of systems began to be integrated into more applications, such as defect detection, metrology, intrusion detection, code reading, and even object and distance detection [3]; being the latter application that will be given more emphasis in this manuscript.

A service robot is a system designed to carry out tasks considered dangerous and repetitive activities or jobs, being partially or totally autonomous [4]. This type of machine requires integrated elements to function optimally in the environment for which it was designed, such as in the case of a navigation or vision system. This type of robot, in turn, can be classified into different groups or categories defined by the performed activity by the service robot [5]. A personal service robot is designed for home maintenance, offering help with housework and even entertaining the family, some of these activities being cleaning floors and windows, vacuuming floors, and even surveillance. By contrast, professional service robots are designed for commercial purposes, such as those used in the medical, marketing, and public cleaning industries [6].

C. E. Mendoza-Ramirez · J. A. Gonzalez-Aguirre · J. C. Tudon-Martinez (✉) ·
L. C. Felix-Herrán · M. A. Ramírez-Moreno · J. D.-J. Lozoya-Santos
School of Engineering and Sciences, Tecnologico de Monterrey, Monterrey, México
e-mail: jc.tudon@tec.mx

O. Sename
GIPSA Lab, Université Grenoble Alpes, Grenoble, France

G. L. Conte and O. Sename (eds.), *Proceedings of 12th International Conference on Mechatronics and Control Engineering*, Lecture Notes in Mechanical Engineering,
https://doi.org/10.1007/978-981-96-6452-8_17

Vision systems in service robots are adapted depending on the situation in which it is used [7]. In the case of the *PiBot* robot presented in [8], there are certain parameters on which the vision system is based to detect or not a person, being these parameters the hardware used (camera and GPU), the neural networks, the database used to enter the networks, the size of the box formed by the neural network (which is responsible for detecting people in the field vision), and the distance from the robot to the target to be detected. As an extension of [8], this paper details the description of each of the elements that make up the vision system, both hardware and software, as well as explains the composition and capabilities of the neural network algorithms used to obtain the accuracy of the results shown in this investigation. The vision system in this research mainly performs on a service robot that detects nearby persons and advises when they don't have a mask or when they don't respect a minimum distance between them. With all the recent COVID-19 problems, *PiBot* is a great option for big public spaces, a campus or a corporation building for example. Also, it can be adapted to different environments or areas without the need for many changes and compared to most robots, it is a cheaper solution [7].

Recently, similar vision systems have demonstrated very good results of face mask detection using different neural networks, but some of these approaches are trained and validated from static images from controlled databases; although these methods are promising solutions because can be implemented for real-time applications [9–11]. However, none of these vision systems have been tested on a service robot whose vehicle dynamics can disturb the video capturing. In this sense, this paper presents the vision system performance of a service robot able to be in a dynamic environment, i.e., the image processing is not static because the robot can be in movement.

The paper is organized as follows: the next section describes the hardware setup and the conformation of its elements through a flowchart in which all the integrated elements are shown and the order in which they interact with each other. Besides, the software setup is also detailed in Sect. 2. Section 3 details the experimentation carried out in this paper and describes the process detection by the *PiBot* vision system, by presenting the parameters considered for an optimal measurement, recommendations, errors, distances, and coordinates, etc. Afterwards, Sect. 4 presents and discuss the detection results through a confusion matrix and, at the end, a conclusion on the performance achieved by the vision system within the *PiBot* service robot is presented in Sect. 5.

2 Experimental Setup

Hardware Setup. The main hardware component of the vision system is the camera [7]. In *PiBot*, an Intel RealSense Depth Camera D435i has been considered, which it has an IMU (Intertial Measurement Unit) with which it is capable of detecting depth under any situation when the camera moves.

The D435i model is designed to be integrated with various software, making it an easy camera to implement within any application or project. In the case of the *PiBot*, the D435i works correctly and easily with the Robot Operating System (ROS) because it is also integrated with this software, being a very versatile component in that aspect. Other software compatible and commonly used with the D435i camera include DOT Product, ImFusion (3D scanning software), Gestoos, Teleidoscope (hand and motion recognition software), and GPC Systems (measurement software), among others.

Due to this vision system works with artificial intelligence, a "digital brain" is required to carry out all the operations and processes necessary to be able to perform different functions. PiBot makes use of an NVIDIA branded Jetson TX2 for this purpose. This CPU has the necessary hardware to work with artificial intelligence and neural networks, designed to carry out tasks with very demanding calculation needs, offering maximum performance levels and relatively low power consumption. The Jetson TX2 has quite robust hardware, hence its great performance capacity, with some components integrated into the board such as a Gigabit Ethernet port, Wi-Fi and Bluetooth connectivity, 8GB of RAM memory, and 32 GB in eMMC format. Other capabilities include working with two 4k video streams at 60 frames per second, and simultaneously managing 6 cameras, and it also has a 64-bit processor and a 256-core Pascal GPU.

Power management of the service robot is through 2 completely identical Turnigy brand 4s Lithium Polymer (Li-Po) batteries (containing 4 cells inside), with a capacity of 14.8 V at 5000 mAh. The batteries are equipped with a discharge cable to minimize resistance and withstand high current loads, as well as having XT-90 and JST-XH connectors. Turnigy 4s Li-Po batteries are the only source of power to power all the components and systems that comprise the *PiBot*, in addition, they are specially designed to be used in vehicles [12], capable of withstanding sudden movements without affecting their performance or put in danger to your environment. Both batteries are connected to a voltage regulator on the bottom of the *PiBot*.

The communication protocol between these components is relatively simple. The Intel RealSense Depth Camera D435i is connected via USB cable to the NVIDIA Jetson TX2. The NVIDIA Jetson TX2 board is in charge of connecting to the network through its Wi-Fi connectivity, in this way it is possible to connect a computer to the *PiBot* as long as the same network is connected. Figure 1 shows that the used camera is located on the front side of the *PiBot*; while the NVIDIA Jetson TX2 board is inside, under the removable cover that is used to protect all the internal components of the robot.

Software Setup. The operation of the vision system detection is based on a rectangle composed of two coordinates which encloses the person who is in the camera field of vision, which is processed by the used software. This process is divided into three phases: detection of the person, cropping of the captured image, and detection of face masks to determine if the person captured is using one or not [8].

The person detection phase is carried out by means of YOLO (You Only Look Once) version 4, an algorithm that uses neural networks specialized in detection by

Fig. 1 *PiBot*, location of the Intel RealSense Depth Camera and NVIDIA Jetson TX2

means of boxes. YOLO was trained using COCO (Common Objects in COntext), a database that was used so that the neural network (in this case YOLO v4) was capable of detecting people through these boxes [13]. In addition to detecting people, the distance is calculated using the D435i camera sensors, in this way it is determined if two people are respecting the minimum distance between them. Based on the flow chart shown in Fig. 2, the input received by the YOLO neural network is given by the images captured by the camera. Once the processing is done by this neural network, the output is received by the Coffee neural network, which crops the body of the image processed by YOLO, in this way it is easier for the third neural network, Darknet, to determine if a mask is being used or not, therefore the Coffee output is the Darknet input, ending the detection process and thus sending an alert in case a mask is not being used.

YOLO is an algorithm that stands out above the others mainly for its speed, precision, and ability to detect several objects in one, being able to detect signals, people, animals, etc. In addition to these advantages, conventional detection algorithms are not capable of detecting objects in a single execution, however, this is possible with YOLO, specifically as of version 3 [13]. Basically, the YOLO algorithm consists of 3 techniques: residual blocks, bounding box regression, and Intersection Over Union (IOU). The residual blocks technique consists of dividing the image into various grids with the same dimensions each. Then, with the bounding box regression, this grid forms a box that highlights an object in an image, having a width and a height (two coordinates). Finally, the IOU technique consists of dividing the area of overlap (the area between the predicted bounding box and the ground-truth bounding box) by the area of union, which represents the area encompassed by both the predicted bounding box and the ground-truth bounding box. If both boxes are the same, the

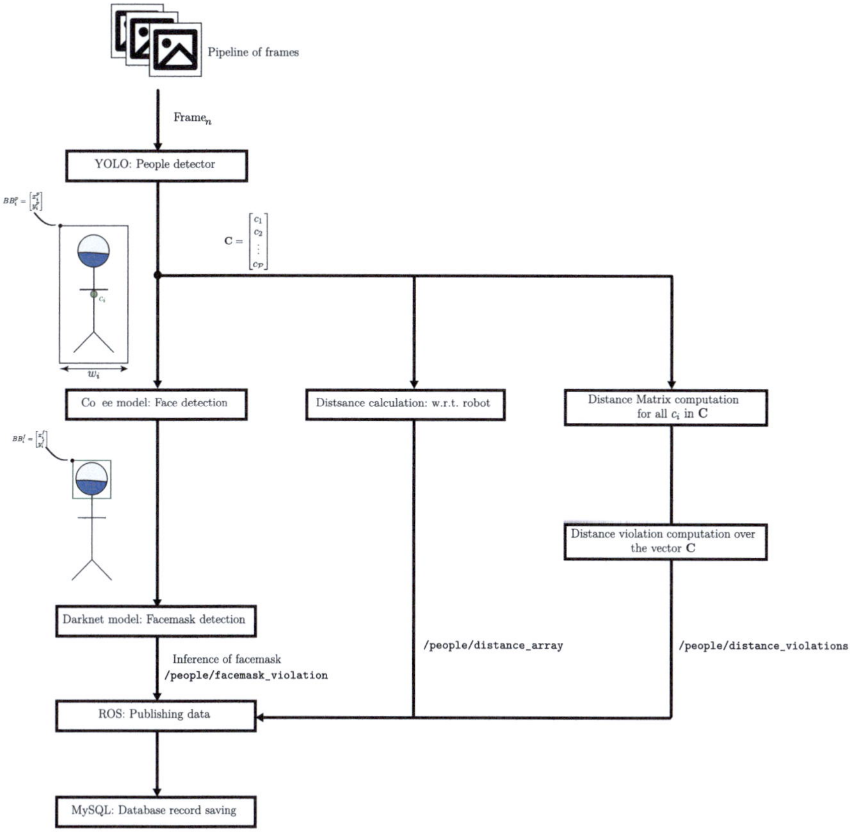

Fig. 2 *PiBot* software structure

IOU ratio is equal to 1, thus the prediction is perfect; if the boxes are not the same, the prediction losses match efficiency [14]. A good prediction could be considered when an IOU is greater than 0.73.

YOLO's performance is superior to that of conventional algorithms, having 48.1% of mean Average Precision (mAP) trained with MS-COCO 2014 database. The frame rate can vary depending on the GPU being used, ranging from 40–90fps with a Titan X, 62fps with a Titan V, and 34fps with an RTX2070 [15].

Darknet is an open-source neural network framework written in C and CUDA. This neural network is specialized in object detection, having a variety of advantages that make it ideal to use, in this case, for object detection. Its easy installation, support for GPUs and CPUs, its compatibility with the YOLO algorithm, and its speed are some of its virtues [16], which make it an excellent option for detecting specific objects in real-time, in this case, face masks. As with YOLO, Darknet performance can vary depending on the hardware being used. With an Intel Xeon CPU at 3.2 GHz,

32 GB RAM, and an NVIDIA GeForce GTX 1060 GPU with 6 GB RAM, Darknet is capable of processing an 86-second video with a total of 4300 frames in a time of 111 s at 38.7 fps [17].

3 Experimentation

Design of experiments. In this experimentation, the controlling factors involved are the selected distances for the two points that will be used in each iteration until the desired degree of precision is reached. This experimentation was designed to be carried out in an uncontrolled environment, that is, uncontrolled factors intervened. In this case, sunlight intensity, people quantity passing by, the distance between people and *PiBot*, and weather were external factors that influenced the samples taken. *PiBot* was placed for 30 min in morning and evening time on the ground floor of the Tecnologico de Monterrey campus Monterrey library to have a panoramic view and take the samples through all the area; the temperature was around 32°C. Since this open area is under a building, relatively little sunlight distorts the results of the samples. During this experimentation, 1,822 samples were taken, which will be analyzed by the vision system, and the results are discussed in Sect. 4 (Fig. 3).

The vision system presents five possible scenarios related to detection. True Positive (TP): The *PiBot* vision system detects the mask when the person is wearing it, True Negative (TN): The mask is not detected because the person is not wearing it, False Positive (FP): Mask is detected but the person does not have it on, False Negative (FN): no mask is detected although the person really does have it on, Miss (MI): the vision system gives this response when there are no people to detect or they are far enough away to not be detected, therefore the system does not make any inference, and finally NA.

The results of the experiments will be:

1. The operating conditions, that is, minimum and maximum distance where the system behaves properly

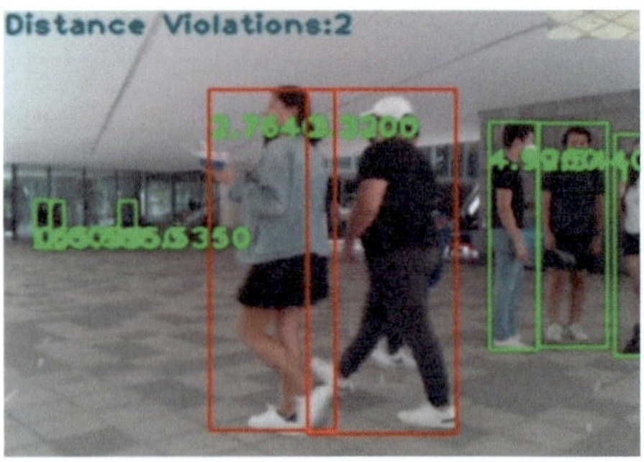

Fig. 3 Tecnologico de Monterrey library, place where samples were taken. In red and green squares are displayed example of bounding boxes, predicted and ground-truth, used by YOLO in the *PiBot* vision system

2. The maximum number of people wearing mask that the robot can detect in a single frame
3. The Accuracy of the system
4. The sensitivity of the system
5. The Receiver Operating Characteristic curve as a function of the threshold of the classifiers

The measured variables will be:

1. The inference of the classifier boolean state of facemask use
2. The number of people in frame
3. Color of the facemask
4. Is the people (on the frame) actually wearing a facemask?
5. Distance with respect to (w.r.t.) the vision camera

The control variable in this experimental work will be the constant distance x w.r.t. the camera of the robot and the measured variable will be the output of the classifier in terms of correct face mask use. These experiments will determine the operating conditions of the classifier module in the robotic system. A distribution of operating conditions will be obtained as output after the experimentation is terminated and the data is processed.

For this experimentation, let's define a function $\Phi : \mathbb{R}^+ \rightarrow \mathbb{N}$ as the number of misses that the classifier does in a given distance x. The data gathering as well as the operating conditions will be searched using a dichotomous search explained as follows. In each iteration, the function $\Phi(x)$ will be evaluated at two points $x_a = x_1 - \epsilon/2$ and $x_b = x_1 + \epsilon/2$ where ϵ happens to be a small arbitrary number. Then, *if* $\Phi(x_a) < \Phi(x_b)$ we center the new interval at x_b, otherwise we center at x_a. Finally, we repeat the process for the next iteration until the required accuracy is reached.

The possible errors that can be obtained in the vision system are defined in Table 1. The precision of the system P is obtained through the Eq. (1) that is determined by the total of each one of the previous types of answers as:

$$P = \frac{fCount[\text{“}TP''']}{fCount[\text{“}TP''] + fCount[\text{“}FP'']} \tag{1}$$

where $fCount$ is the program command that stores the cumulative total of each type of response (TP and FP) in this case.

Table 1 Types of error in a hypothesis test

Name	Symbol	Description
True Positive	TP	The predictor hits the expected positive result
True Negative	TN	The predictor hits the expected negative result
False Positive	FP	A false alarm of a negative (Type I Error)
False Negative	FN	A miss of the predictor on a positive result (Type II Error)

4 Vision System Results

Processing of samples. Once the samples were taken, two histograms and a density histogram were made to represent the results obtained. The density histogram shown in Fig. 4 illustrates a top view of the TEC campus Monterrey library, in which the distribution of the number of samples taken is represented by means of colors, yellow being the place where a greater number of samples were taken and the red tones where it was not possible to obtain a large number of them.

Figure 5 shows the histogram distance versus mask violation inference probability, being the distance from the camera to the objective of the sample taken in meters, and the probability is represented between a range of 0–1. These results show that the optimal distance to take the samples is between the range of 2–2.5 m, reaching a probability of mask violation inference of 71%. In minors or majors ranges, this probability is negatively affected, with the most deficient ranges being those between 0 and 0.5 m and 3 to 3.5 m.

The time that people were exposed to the vision system influenced the probability of mask violation inference of the samples taken. The ideal time range is between 20 and 40 s since this range belongs to the 71% probability of a good mask violation inference. These data can be seen in more detail in Fig. 6.

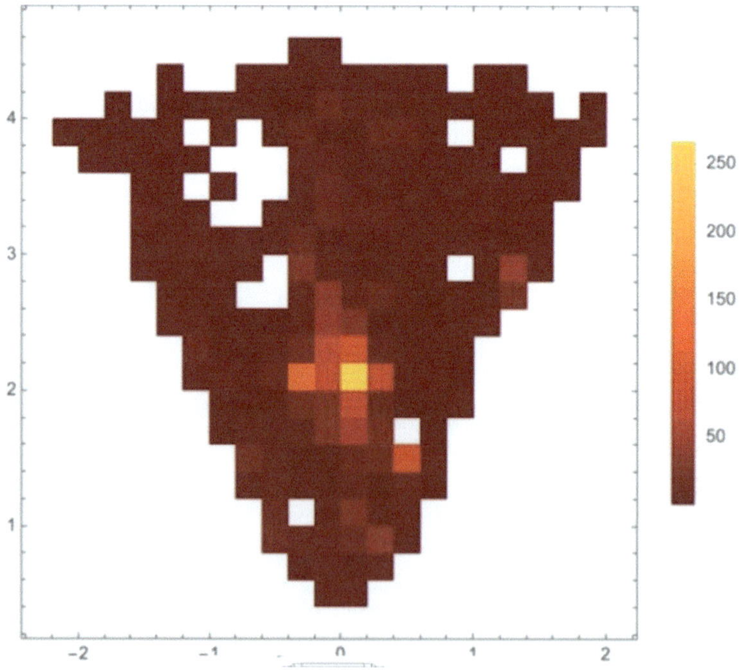

Fig. 4 Density histogram of the total samples taken

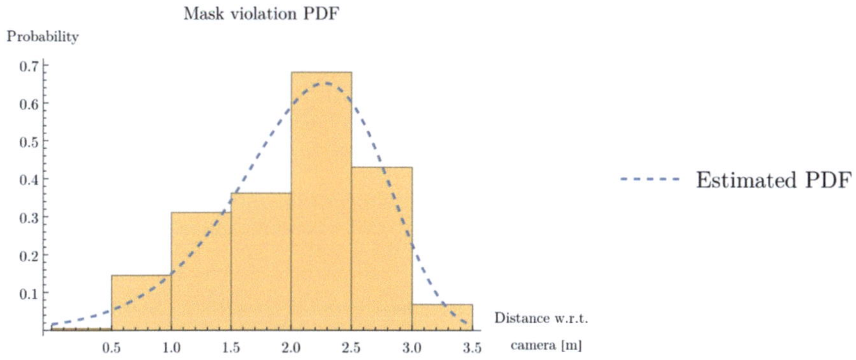

Fig. 5 Good mask violation inference probability versus distance

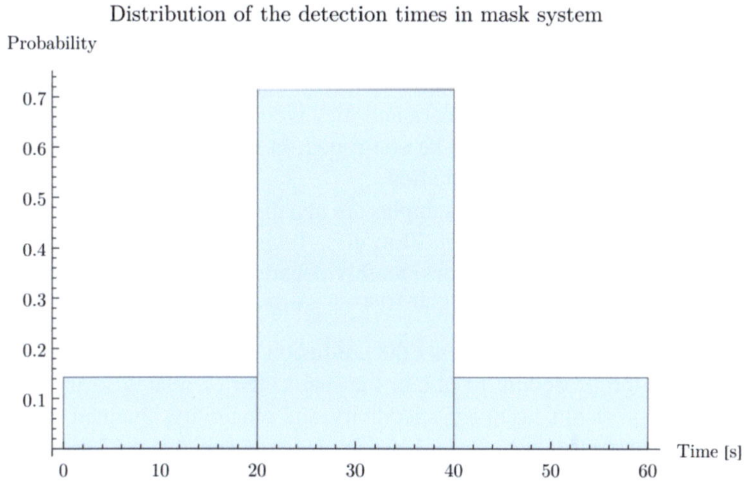

Fig. 6 Distribution of the inference time of the mask detector system

The classifier to process the samples taken was developed using the logistic regression method, using the Wolfram Mathematica software to write the code and be able to generate the above image processing results.

Some possible solutions that could improve the sampling on the vision system are: (1) classify the video samples by some feature extraction to separate true positives, false positives, etc., and calculate a confusion matrix based on these data; (2) correlate the attention ratio to the probability of a good detection; (3) correlate relative luminosity to the probability of a good inference; and (4) generate a Poisson process model to estimate the number of people at any time.

Precision. The maximum detection accuracy obtained for the vision system using the trained neural networks, hardware, software, and methods mentioned in previous sections is 71% under real and uncontrolled conditions, this means, the people were

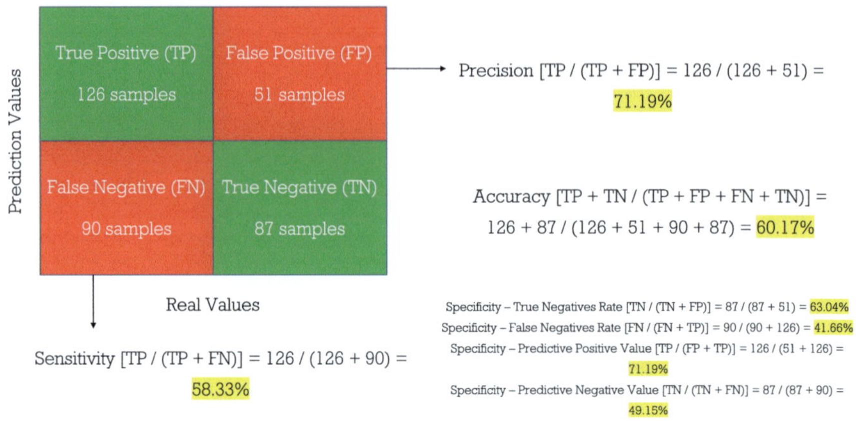

Fig. 7 PiBot vision system confusion matrix and metrics

not standing up in front of the camera and they were not at a constant distance, as a conventional database that could be controlled. In fact, the pedestrian traffic also was uncontrolled in this experimentation.

The ideal conditions for taking samples are at a distance of 2–2.5 m and an exposure time in front of the camera of 20 to 40 s. Taking samples in lower or higher ranges of distance and time can considerably negatively affect the accuracy of the PiBot's vision system, reaching below 10% on some occasions.

Confusion Matrix. Figure 7 presents the confusion matrix obtained in this experimentation and the procedure to obtain the main metrics related to the detection results, such as precision, accuracy, specificity, and sensitivity, obtained through the samples taken, being true positive = 126 samples, true negative = 87 samples, false positive = 51 samples, false negative = 90 samples, misses = 235, and NA = 1233. NA means basically those images captured by the *PiBot* automatically in a sampling time, but the image does not capture any person.

The obtained data from the previous confusion matrix shows an accuracy of 60.17%, sensitivity of 58.33%, and specificity of 63.04%, 41.66%, 71.19%, and 49.15% (True negatives rate, false negatives rate, predictive positive value, predictive negative value respectively).

5 Conclusion

This paper presents the detection results of the vision system performed on a service robot, whose architecture has been designed as a cheaper solution than commercial devices. Manufacture details of this robot, called *PiBot* can be reviewed in [8].

Although the detection results show 71% accuracy in this experimentation (under uncontrolled factors such as people quantity passing by, the distance between people, the perspective of captured faces, etc.), these results can be interpreted as preliminary, demonstrating that this own-designed and manufactured service robot has a good potential to achieve excellent results on its vision system The major opportunity area is the improvement of the training algorithm using a better database available for this purpose.

References

1. Huang TS (1996) Computer vision: evolution and promise, pp 21–25
2. Li D, Dong Y (2014) Deep learning: methods and applications. Assoc Comput Mach 7:3–4
3. Szeliski R (2022) Computer vision: algorithms and applications. Springer Nature
4. Schmidlin EA, Jones KS (2011) Human-robot interaction: toward usable personal service robots. Reviews of Human Factors and Ergonomics
5. Chiang A-H, Trimi S (2020) Impacts of service robots on service quality. Service business
6. Iborra A et al (2009) Design of service robots. IEEE Robot Autom Mag
7. Gonzalez Aguirre JA et al (2021) Service robots: trends and technology. Appl Sci 11(22)
8. Osorio-Oliveros Ricardo et al (2021) PiBOT: design and development of a mobile robotic platform for COVID-19 response. Adv Autom Robot Res 347:252–260
9. Nagrath P et al (2021) SSDMNV2: A real time DNN-based face mask detection system using single shot multibox detector and MobileNetV2. Sustain Cities Soc 66
10. Razavi M et al (2021) An automatic system to monitor the physical distance and face mask wearing of construction workers in COVID-19 pandemic. SN Comput Sci
11. Akhil Kumar et al. "A hybrid tiny YOLO v4-SPP module based improved face mask detection vision system". In: *Journal of Ambient Intelligence and Humanized Computing* (2021)
12. Chang K et al (2016) LiPo battery energy studies for improved flight performance of unmanned aerial systems. Unmanned Systems Technology XVIII
13. Ma Y et al (2020) Object detection through modified YOLO neural network. Sci Program
14. Ergu Peiyuan Jiang Daji et al (2022) A review of yolo algorithm developments. Proc Comput Sci 199:1066–1073
15. Feng H et al (2022) Benchmark analysis of YOLO performance on edge intelligence devices. Cryptography 6(2)
16. Koo Y, You C, Kim S (2018) OpenCL-darknet: an openCL implementation for object detection. Big Data and Smart Computing
17. Koo Y, Kim S, Ha Y-G (2020) OpenCL-darknet: implementation and optimization of OpenCL-based deep learning object detection framework. Artif Intell Big Data Comput

Lean Manufacturing and Industrial Automation

Warehouse Efficiency Improvement Model Applying Lean Manufacturing Methodology in a Roller Shades Company

Adrián Humberto Cuenca Bazán, César Augusto Arrunátegui Vásquez, and José Antonio Taquía Gutiérrez

1 Introduction

Currently, there is a high level of competitiveness among different industries, where the warehouse plays an essential role in the supply chain of companies. As the idea of developing excellent customer service grows, organizations see the need to generate warehouses with a more efficient flow, being able to protect themselves from variations in demand levels [1]. In this sense, the term "warehouse" is a physical space in which certain types of products and/or materials are received, transferred, sorted, and shipped from one point to another [2].

Aligned with this constant increase in the need for an efficient warehouse flow, different authors expose the different recurring problems that delay and/or affect the development of this objective, which can be linked to the structure and placement of the warehouse [3] as well as in the timing of the realization of the storage process [4].

However, making the necessary changes to adapt the warehouse to a new workflow may generate certain delays due to its daily necessity within the normal processes of organizations. In this sense, simulation is a viable option to solve these implementation problems, capturing the problems and possible solutions virtually, and being able to generate better decision-making related to storage [5].

Therefore, considering the problems identified in the analyzed company, the following research question is proposed: What solutions can be implemented to increase the efficiency of the storage process? The objective of this article is to seek viable solutions through the principles of Lean methodology to the occurrence of these problems within the company under study, achieving an increase in efficiency,

A. H. C. Bazán · C. A. A. Vásquez · J. A. T. Gutiérrez (✉)
Universidad de Lima, Lima, Perú
e-mail: jtaquia@ulima.edu.pe

G. L. Conte and O. Sename (eds.), *Proceedings of 12th International Conference on Mechatronics and Control Engineering*, Lecture Notes in Mechanical Engineering,
https://doi.org/10.1007/978-981-96-6452-8_18

and in the regular flow of work, in the storage process, as well as its corresponding validation through simulated replicas of the selected process. In this way, this work will contribute to both researchers and professionals interested in the internal logistics of companies in the textile sector.

This will be followed by a review of the state of the art. Then, the theoretical framework will be developed, which will explain concepts related to the warehouse and the different ways to solve the existing problems. Subsequently, the experimentation will be developed based on the proposed methodology. Finally, the results obtained will be discussed and the conclusions of this research will be established.

1.1 Lean Manufacturing Methodology Applications

Identifying the main problem as the long duration of the production process and the long transports and movements of materials, since they have a traditional production system, they make use of a study of movements, a standardized work through the taking of time and a balance of lines, achieving a reduction of cycle time and lead time of 32.73% and 11.8%, respectively, and an increase in productivity of 16.66%. Highlighting the time study, this tool fulfilled the task of collecting the duration of the different tasks of the production process, selecting a single type of product as a baseline [6].

On the other hand, [7] embodies the use of the 5S tool, the study of the work, and the elimination of the mudas to attack the low productivity within the storage area. Detailing the use of the 5S tool, a new warehouse area was established which generated an easier workflow for the operators and an increase in the quality of the final product; likewise, there was an increase in the percentage of compliance with the standards established for 5S to 66.67%.

1.2 Plan–Do–Check–Act Methodology Applications

The work developed [8] makes use of the PDCA (Plan, Do, Check, Act) methodology, which is a tool for continuous process improvement. This methodology is complemented by other techniques such as Lean Warehousing, Slotting, Wave Picking, and 5S. Thanks to the implementation of these methodologies, a 19.12% reduction in process time was achieved. In addition, the distance traveled in the warehouse was reduced from 244.5 to 173.5 m. The focus of the article is to highlight the benefits and results obtained through the implementation of the PDCA methodology and the 5S organizational tool, these approaches have proven to be effective in improving efficiency and productivity in the warehouse.

The PDCA methodology together with the 5S, Jidoka, Poka-Yoke, and Standardized Work tools. Thanks to these tools, an increase of 15.98% in the perfect order ratio and a reduction of 19.95% in order preparation time were achieved [9]. These results demonstrate that the application of Lean Manufacturing tools, combined with the PDCA approach, has been successful in optimizing processes in the freight distribution company. In addition to improving the quality of service, it was also possible to observe an improvement in the organizational culture within the company.

On the other hand, used the PDCA methodology with the 5S, Kanban, and Systematic Layout Planning tools, which were applied to solve problems such as the accumulation of damaged items in the warehouse, unproductive times due to unnecessary movements, late delivery of orders to the client and reprocessing due to inadequate work procedures [10]. The results obtained were an improvement in productivity of 24.99% and an economic benefit of PEN 66 112 per year (Peruvian currency), which indicates that the implementation of Lean Manufacturing and the tools mentioned were successful in solving the problems identified, achieving a considerable increase in the efficiency and profitability of the automation company.

2 Methodology

In the company's constant search for improvement and innovation, it was decided to focus the research on the storage of the Roller Shades product. In this sense, for the development of this work, the Lean Manufacturing methodology and the Arena simulation software have been developed for the corresponding validation.

As a general framework, the research methodology will consist of the 4 phases of the PDCA methodology (Fig. 1) and 2 sub-phases which are: the development of the 5S tool and the validation of the Arena simulator pilot test.

In the first "Plan" phase, an exhaustive evaluation of the current state of the storage area was carried out, identifying the main problem and its root causes to support the implementation of the pilot test. In addition, an awareness-raising talk

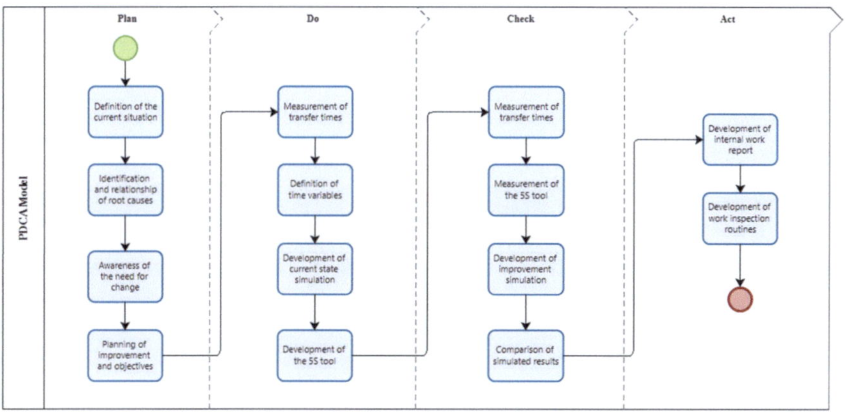

Fig. 1 PDCA methodology development model

was organized to highlight the current situation of the selected area and the benefits that would result from the implementation of this pilot test. The improvement strategy was clearly defined, laying the foundations for the effectiveness of the improvement process, and the objective of increasing efficiency by 10% was established.

For the "Do" phase, the initial times of the storage process were measured, as well as the respective statistical variables, for the development of the simulation in the Arena software. Subsequently, the status of the area designated for the 5S tool was measured and its implementation was started for the pilot test.

For the "Verify" phase, the storage process times were measured according to the data obtained from the pilot test and the respective 5S evaluation to make a comparison concerning the initial state. Subsequently, the simulation was carried out with the information from the pilot test to validate whether the improvement idea was feasible.

Finally, for the "Act" phase, an internal report was developed that shows the correct development of the work in the storage area and a series of inspection routines to validate the process.

The sub-phase of the development of the 5S tool consists of a pilot test in a specific space in the storage area, which will serve as a model to feed, with the data collected, the Arena software and be able to perform the corresponding simulation and compare the results with the current transfer times that the company has. In the same way, the Arena software will be used in 2 key points of the research: the simulation of the current state of the warehouse and the simulation of the pilot test, to validate the improvement that could be generated in the warehouse.

3 Experimentation

For process improvement validation we prepared a simulation model as seen in Fig. 2. Within the manufacturing process of the Roller Shades, once the manufacturing stage of each shade is completed, they are grouped into sets of three, classified according to the type of product to which they belong.

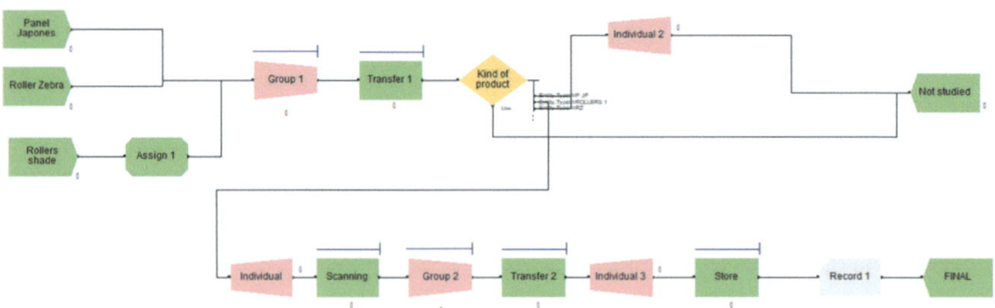

Fig. 2 Simulation in Arena software

After the groups are formed, the operators take them to designated tables to scan the bar codes; each vertical curtain has a unique bar code containing detailed information on its characteristics and specifications (during the time analysis these operators will be referred to as "operator 1"). The operators proceed to scan the barcodes of the curtains in each group; this scanning facilitates quick and accurate identification of the curtains during storage and subsequent shipment (during the time analysis these operators will be referred to as "operators 2").

Once the scanning is completed, the groups of three curtains are regrouped again for further transfer, these are taken to the final warehouse by the operators (during the time analysis these operators will be referred to as "operators 3"). The last process consists of arranging each product individually (during the time analysis these operators will be referred to as "operators 4").

3.1 Arena Software—Pre-improvement Simulation

For the time study before the improvement, the most experienced operators within the company were selected and it was verified that the machinery used during the process was in optimal conditions; likewise, a stopwatch and an Excel sheet were used where the values obtained by the operators were put.

Two-time studies were carried out, one before the pilot test and the other during the pilot test. A sample of 100 times was taken which were seen in each process: Operator 1, Operator 2, Operator 3, Operator 4.

Having the data of the respective times by operators, we proceeded to use the "Input Analyzer" tool, which allowed us to obtain the correct distribution using Arena software. Similarly, the percentage error of the graph will be taken into account: the Chi-Square test and the K-S test; according to [11], if the "p-value" obtained in the Chi-Square test is greater than $0.05 = 5\%$ the value correctly adjusted, likewise with the K-S test, if the "p-value" obtained is greater than 0.15, the value is correctly adjusted.

Continuing with the time study, a sampling of 15 simulations was carried out in the Arena software obtaining the following statistical results.

To find the sample size we used the t-statistic, through the formula DISTR.T.INV (0.05;14). The value 0.05 represents the probability and the value 14 is the degrees of freedom which is "n − 1". The Result can be seen in Table 1.

Table 1 Sample size definition

Define sample size		
Stand. dev	2.651	
t	2.14	
e	1.038	minutes
n	**30.0**	**31**

Table 2 Average units of outflows in the storage process

	Average units
Number of outputs	36
Number of products entering the process	106

At the time of executing the above formula, it generates a standard error of 1.468 min, which is interpreted with 95% confidence with an error of 88.09 s concerning the sample. Having the standard error, the sample size is defined with the following formula, which is interpreted as "number of samples is equal to the result obtained by squaring the multiplication of the standard deviation by the t-statistic, divided by the standard error." The t-statistic value is the same as the one found in the standard error estimate.

$$n = \left(\frac{\sigma t}{e}\right)^2 \tag{1}$$

When estimating a new standard error, the sample number changes to 31. With these new parameters, the Arena system was run and obtained Table 2 values.

According to the data obtained, the total number of products that finish the whole process on average is 36 units. This indicates that the efficiency is 33.96% of the whole storage process.

$$Efficiency = \left(\frac{Number\ of\ outputs}{Number\ of\ inputs}\right) * 100 \tag{2}$$

$$Efficiency = \left(\frac{36}{106}\right) * 100 = 33.96\%$$

In the same way, the values of waiting time and total time of the storage process at the end of the Arena software execution are obtained.

3.2 Pilot Test—5S Tool

Before the development of the 5S tool, training was provided to employees involved in warehouse activities so that they could understand the advantages of the constant and conscious use of this tool and the positive effects generated by its use.

For the 1°S Organize (Seiri), the labeling of the sliding curtain shelves was developed to differentiate the workspace as a pilot test. For 2°S Tidy (Seiton), stickers were used on the floors to allow signage of staff areas. For 3°S Clean (Seiso), daily cleaning routines were established, before and after the start of routine activities, with the assignments of the collaborators within the warehouse who will carry out this activity.

Fig. 3 5'S evaluation

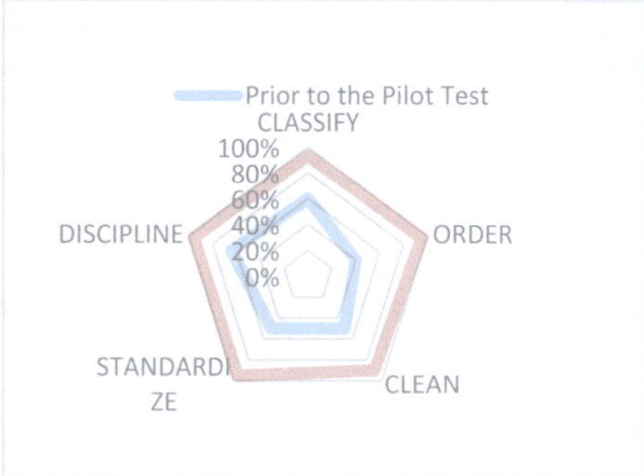

To contrast the workflow obtained with the flow before the pilot test, a checklist was developed to identify, in percentage terms, the improvement achieved with the 5S tool.

Similarly, for the 4°S Standardize, a comprehensive report has been prepared on the work performed correctly, as well as guidelines for maintaining the optimal state of the storage area. This report replicates the results obtained in the pilot test and will serve as a guide for employees in the future.

Finally, for the development of the 5S Discipline, inspection routines were implemented both in the storage area and the corresponding process, carried out by the operators. The objective is to reinforce and encourage the use of the 5S methodology so that it becomes a sustainable habit over time. In this way, the aim is to guarantee efficient and orderly management that benefits both the employees and the storage area in general.

To compare the values before and after the development of the 5S tool, a radial graph was created that shows the values of each "S" for each scenario, as well as the corresponding variation between the two (Fig. 3).

Similarly, to measure the significant improvement resulting from the implementation of the 5S methodology, a questionnaire was sent to employees to evaluate their level of satisfaction. Using a scale of 1–5, where 1 means "no improvement" and 5 means "very significant improvement", the operators expressed their agreement. According to the results obtained, 80% of the operators indicated that they experienced a very significant improvement, while the remaining 20% indicated a significant improvement.

3.3 *Arena Software—Post-improvement Simulation*

Similarly, no changes were made to the structure of the preliminary simulation commands since the storage process has not been changed.

According to the information presented, there is not enough evidence to reject the null hypothesis that the data examined, after the pilot test, adhere to a Beta distribution.

When simulating Arena software with the new distribution data for the storage steps performed by the operators, an increase in the number of finished product outputs was obtained. According to the data obtained, the total number of products that complete the entire process on average is 58 units, which indicates an increase in efficiency to 54.72% of the entire storage process.

$$Efficiency = \left(\frac{58}{106}\right) * 100 = 54.72\% \quad (3)$$

4 Results and Discussion

A visual comparison shows a considerable improvement in the storage process of the sliding curtains; specifically, there was a decrease of more than 80% and 65% in the waiting times of the storage processes and the total time of the overall storage process, respectively. Likewise, there was an increase in efficiency to 54.72% in the storage process (Table 3).

In the same way, to verify the theory, the "Output Analyzer" was used and both simulations were compared (Fig. 4), obtaining the following values. Results in Table 4.

Making a comparison with different research articles, such as the researches of [7] and [10], a similarity with the results obtained by the development of the pilot test and the simulation can be evidenced, being these 94% and 20.76% in the evaluation of 5'S and the increase of efficiency, respectively, of the research carried out, and 66.67% and 24.99% the respective values of the selected articles.

Table 3 Comparison of the efficiency of the storage process

	Before pilot test (%)	Post pilot test (%)	Variation (%)
Efficiency	33.96	54.72	61.13

Fig. 4 Comparison of the average times in output analyzer

Table 4 Comparison of simulations in output analyzer

Identifier	Average time	Standard deviation	Half width of 95% confidence interval	Minimum value for the mean	Maximum value for the mean	Number of observations
Simulation 1	141	7.62	2.8	126	161	31
Simulation 2	47	0.663	0.243	45.7	48.3	31
V1 versus V2	94.2	7.51	2.76			31

Similarly, in both [5] and [9] research, there is a similarity with the results generated by the Arena software, being the results of 55% in loading and unloading times and 50% in receiving, storage, and picking times, respectively, for the research articles and 66.71% of the present study. These findings are highly promising, as they demonstrate the effectiveness of Arena software in optimizing logistics processes and reducing times; by implementing this approach in the warehousing area, substantial improvements in efficiency and productivity can be achieved, which in turn has a positive impact on operating costs and response time in the delivery of materials.

5 Conclusions

While the company may have a system of order in place, it is important to develop a visual plan and carry out awareness-raising to ensure an efficient implementation of the 5S methodology compared to the initial state. Through this methodology, we were able to verify that the commitment of the entire organization, from top management to operators, is fundamental to achieving significant improvements in the company.

If the data is not properly configured, there could be errors when using different simulation software available on the market. Therefore, it is crucial to make sure that the parameters are accurate and representative of reality to obtain reliable results and make informed decisions. In the same way, the lean manufacturing methodology provides a structured work framework, through the application of a set of tools, which facilitates the adoption of improvement proposals that are affordable for the company and innovative and easy to understand for the collaborators. However, it is necessary to train the collaborators in these tools and methodologies for their internalization and maintenance over time.

References

1. Zahra F, Moussa B, De Guio R, Dubois S, Rasovska I (2019) Computers & industrial engineering study of an innovative method based on complementarity between ARIZ, lean management and discrete event simulation for solving warehousing problems. Comput Ind Eng 132(April):124–140. https://doi.org/10.1016/j.cie.2019.04.024
2. Cakmak E, Gunay NS, Aybakan G, Tanyas M (2012) Determining the size and design of flow type and U-type warehouses. Procedia Soc Behav Sci 58:1425–1433. https://doi.org/10.1016/j.sbspro.2012.09.1127
3. Palange A, Dhatrak P (2021) Materials today: proceedings lean manufacturing a vital tool to enhance productivity in manufacturing. Mater Today Proc 46:729–736. https://doi.org/10.1016/j.matpr.2020.12.193
4. Muñoz-Villamizar A, Solano-Charris EL, AzadDisfany M, Reyes-Rubiano L (2021) Study of urban-traffic congestion based on google maps API: the case of Boston. IFAC-PapersOnLine 54(1):211–216. https://doi.org/10.1016/j.ifacol.2021.08.079
5. Emami SB, Arabzad SM, Sajjadi SM (2014) A simulation study on warehouse loading system: the case of poultry feed production factory. Int J Logist Syst Manag 19(3):347. https://doi.org/10.1504/ijlsm.2014.065501
6. Mulugeta L (2020) Productivity improvement through lean manufacturing tools in Ethiopian garment manufacturing company. Mater Today: Proc 37(Part 2):1432–1436. https://doi.org/10.1016/j.matpr.2020.06.599
7. Burawat P, (2019) R. U. of T. T./P. Faculty of business administration Thailand, Productivity improvement of corrugated carton industry by implementation of continuous improvement, 5s, work study, and Muda elimination: a case study of xyz Co., Ltd. Int J Eng Adv Technol 8(5c):178–183. https://doi.org/10.35940/ijeat.e1026.0585c19
8. Gómez-Cárdenas MF, de la Cruz-Felipe CP, Felipe-Bravo GM (2021) Proposal for the implementation of lean manufacturing tools in an automation company, Trujillo-Peru, 2020. Propuesta de implementación de Herramientas de Lean Manufacturing en una empresa de automatización, Trujillo-Perú, 2020. In: Proceedings of the LACCEI international multi-conference for engineering, education and technology, vol 2021. https://doi.org/10.18687/LACCEI2021.1.1.194
9. Zelada-Muñoz LA, Arana-Grijalva AC, Quiroz-Flores JC, Cabrera-Gil-Grados EM (2022) Operations model based on lean manufacturing to increase the perfect order rate in SMEs in the consolidated freight distribution sector. In: Proceedings of the 8th international conference on industrial and business engineering
10. Gómez-Cárdenas MF, De la Cruz-Felipe CP, Felipe-Bravo GM (2021) Propuesta De Implementación De Herramientas De Lean Manufacturing En Una Empresa De Automatización, Trujillo-Perú, 2020. In: Proceedings of the 19th LACCEI international multi-conference engineering, education, and technology: "Prospective trends technology skills for sustainable social development "Leveraging emerging technology to construct the future
11. Torres Vega PJ (2016) Simulación de sistemas con el software Arena. Universidad de Lima. Fondo editorial

A Human-Centric Assembly 4.0 System–An Evaluation from the Human Perspective

Jacqueline Humphries, Pepijn Van de Ven, and Alan Ryan

1 Introduction

Industry 4.0 has brought about increased complexity in manufacturing [1]. This is due to many factors including market demand toward customized manufacturing. There is also an expectation to adapt quickly and with real-time production planning and control. However, increasing complexity in manufacturing results in increasingly complex manufacturing processes, which present growing cognitive demands on the operators who assemble the products [2]. Operators have been substituted by automation on high-volume routine assembly and moved to low volume, more complex, and flexible assemblies. The dynamic nature of flexible assemblies requires locating, understanding, and following a wider range of assembly instructions, captured in the Standard Operating Procedures (SOP). Yet the instructions on the SOP are often ignored by operators, who rely on their own experience to perform the assembly process [1]. The increasing complexity presents increasing challenges in relation to production validation and quality checking [3]. There are different approaches to handle this. In one multi-national Medical Device Manufacturer with a base in Ireland, due to the high opportunity for error on some product lines, known as Triple A products (AAA), one Quality Inspector stands between two operators to double check all work as it is being performed. For every AAA product saved there's a scrappage saving of €15,000, in addition to the further savings that could be made to product cycle time, work in progress, and the quality function. However, having

J. Humphries (✉)
Technological University of the Shannon, Limerick, Ireland
e-mail: jacqueline.humphries@tus.ie

J. Humphries · P. Van de Ven · A. Ryan
University of Limerick, Limerick, Ireland

J. Humphries · P. Van de Ven
Confirm Manufacturing Research Centre, Limerick, Ireland

G. L. Conte and O. Sename (eds.), *Proceedings of 12th International Conference on Mechatronics and Control Engineering*, Lecture Notes in Mechanical Engineering,
https://doi.org/10.1007/978-981-96-6452-8_19

a quality engineer standing over the operator's shoulder checking each step of work is not conducive to a pleasant work environment, and thus leads to higher turnover. The turnover of production operators with the associated increased training costs and inconsistent production also leads to poor quality. According to the Head of Innovation, Research, and Development in the same Medical Device Manufacturer, a typical line might employ ten production operators with an average annual turnover of 17.9% . The real cost of employee turnover can be anywhere from thirty percent up to five hundred percent of the position's annual compensation. A new operator reaches full productivity at month three, but it may take up to month six. During such time throughput is reduced, the number of units failing quality checking is higher, and the effort required on the supervision function is higher—all of which have a cost involved. Thus there is significant waste and a significant quality control problem. Another Medical Device Manufacturer in Ireland uses six different post-production quality control checks at different stages in the factory. Often these are destructive acceptance sampling, which within this sector could have a yield loss of 2.5% of the manufactured product [4, p 3]. Any reduction in this loss percentage by either a change in how the products are checked for quality or reducing the number of failed products being produced is more efficient, a considerable saving for the company, and more sustainable with respect to waste.

Assembly Systems (AS) provide assistance to the operator on the assembly process, traditionally by displaying the instructions on a screen in text and graphic form. AS are effective to support high-quality assembly outputs for users of varied experience and ability [5]. There is the potential for the digital transformation that enabled Industry 4.0 to be applied to AS, in a concept known as Assembly Systems 4.0 (AS4.0).

There are a number of theoretical and conceptual studies on the subject of Assembly 4.0, which include a roadmap [6], concepts [7, 8], and a high-level architecture for such a system [9]. However, more studies on the validation of AS4.0 using empirical research methods are required to increase the rate of AS4.0 adoption in complex manufacturing. In particular, given that the success of AS4.0 adoption is dependent upon the technology acceptance by the workers, the lack of empirical evidence from the human perspective is a concern [10]. The research question addressed herein is: Will an Assembly 4.0 System be acceptable to the workers' using the system? The contribution of this paper is an evaluation of an AS4.0 system which is completed from the workers' perspective.

The rest of this paper is structured as follows: first a review of related literature is presented. The next part of the paper contains the main case study, in which an exemplar AS4.0 system is described and applied to a use case. The case study firstly concentrates on a description of the system, followed by an outline of the experiments used to evaluate the system. Finally, the results of the case study, discussion points, and future plans are summarized in the conclusions.

2 Review of Related Literature

The concept for a basic high-level architecture for AS is outlined in Quint et al. [9]. The architecture should contain a Messaging Server, A Workflow Model, Triggers, and Views. Once a process is identified, a messaging sensor loads the workflow model, and the first task in the process is communicated to the operator. This can be done through a monitor display or augmented reality projections of animations or text instructions provided on the workstation. Highlighting boxes of parts or appropriate tools through a pick-by-light type sensor can be integrated. A depth camera can be used to identify the activity of the operator, for example, to identify whether a hand reaches into a box of parts. This can be integrated with other sensors such as a button press. Once a trigger is received, the task state is communicated and the next task in the process is loaded. The next task is made visible to the operator. All AS follow this basic workflow, albeit different interfaces exist.

In Rupprecht et al. [8] a proof-of-concept (POC) assembly system is evaluated. The goal of the POC and the experiment is to evaluate whether operator-picking errors (performance) and the dead time between picks (productivity) are reduced. The authors claim that the use of the system may result in additional benefits of reducing the workload on the employee and enabling less experienced operators to perform the process efficiently and safely. As in Quint et al. [9], a pick-by-local-light (PbLL) system is used. The modules are directly mounted on the shelf, the quantity to be picked is shown on a display, and lights installed in the module indicate the picking position. The operator acknowledges that the task is complete by the use of a push button. While the system shows promise, challenges to the modularity of the system for wider system integration are reported. Research gaps identified are the need for further empirical evidence and field tests, an evaluation of the reaction time of the system, stresses that occur for the assembly staff, and the robustness of the assistance system. In Bertram et al. [7] similar research considers accurate detection of picking actions. They also report the need for further research to consider systems integration. They identify a need to develop a better understanding of what the operator is doing and provide appropriate assistance to the operator in order to enable a more user-centric implementation of assembly systems.

The need for context awareness to determine the task the operator is performing and provide appropriate help for that task is a research gap [7, 11]. An attempt to address the awareness of context is given by Bertram et al. [7], who use Human Activity Recognition (HAR) in the assembly system. The system watches the actions of the operator and checks that the actions correspond with an a priori determination of how that work should be done. In their system, the specifications for the task are visualized, and an animated picture of the step is displayed. A depth imaging sensor can identify the actions of the operator such as picking up a tool. The workflow processor checks the identified actions and matches them with the actual task. Context-specific assistance can be offered. If the appropriate actions are not identified, additional information or hints are presented. Once that task is complete, the system provides information on the following task to be completed. While interesting from a technical perspective, there is a fundamental misalignment with the notion of Operator 4.0. Central to the notion of Operator 4.0 is that the operator is a skilled operator who relies on their own gut feeling, intuition, and experience in the execution of their work. It is estimated that this occurs in about 80% of assembly tasks [12]. Using HAR may result in a dumbing down of work, which ignores the experience of the operator. Not only is the operator being directed on what to do with respect to the process, but with HAR is directed exactly how they must do it. Each person is unique, from the dominant hand through to how they perceive and interpret instructions, and as such would require a personalized HAR system for every process step they work on, in a system that has somehow managed to capture the operators' experience, tacit knowledge, gut feeling, and skill-level. Further, it is known that there is a mismatch between assembly instructions and the way they are actually used by operators in practice [13]. HAR systems developed around how the designer believes assembly instructions will be used will be checking for the wrong action. Either the HAR system is very basic with a limited number of actions such as the pick and twist action, or the level of personalization required is too onerous.

Evaluation studies of AS4.0 systems are largely absent from the literature, however, there are several studies that compare the effectiveness of different interfaces for the delivery of instructions, such as the use of two-dimensional drawings and textual instructions [14], audio [5], Mixed Realities [5] and Augmented Reality (AR) [15, 16], and the information carrier to be used such as screen or tablet [11, 17].

In Havard et al. [16] a comparison of delivery of operator instructions via AR is made to instructions via text and image displayed on a PDF reader. While the use-case is the delivery of maintenance instructions, not assembly, this study is relevant in that step-by-step instructions are provided to an operator. It is reported that the mean duration for carrying out the task does not depend on the device used, that is, the AR does not facilitate faster work. Cycle time is considered from two perspectives: time spent on execution and time spent on consultancy, that is locating and consulting the instructions. They evaluate with twenty students—ten evaluate the system having used the PDF instructions, and ten having used the AR. Usability is evaluated using the System Usability Scale (SUS). The average SUS scores on the PDF instructions were 87.1 by beginner-intermediate operators, and 88.3 by advanced operators rendering it "excellent" in terms of its usability; with an overall average of 87.7. However, with respect to the AR device, this drops to 78.3 "good" by beginner-intermediate operators and 91.2 "excellent" by advanced operators; an overall average of 84.75. They find a 34% reduction in consultation time with the AR compared to the PDF record, only when the AR search time is removed. This shows that the operators more rapidly understand the process with AR, once the instructions have been presented. They find that this diminishes with experience. Beginning or intermediate operators take longer than expert operators who understand globally what is expected of them. Interestingly for a simple task, it was quicker to use the PDF instructions than the AR instructions. With the challenges of using AR in assembly with respect to fatigue, bad lighting, and other factors

[5, 8], it might therefore be more beneficial not to use AR, but to simplify all tasks and in doing so reduce the cognitive load required to understand the simple instructions.

In Pacaux-Lemoine et al. [11] alternative human-machine interfaces, designed in line with principles stemming from the Human-Machine Cooperation approach, are evaluated. Nine participants perform three tests for up to a two-hour period each. The participants score the interfaces on readability, comprehensibility, and relevance of the explanations using a seven-point Likert scale. Secondly, participants' situation awareness is assessed with respect to the process and the information provided by the assistance system. Finally, error detection and error correction performance are assessed. The more explanation the assistance system provides about its analysis of the process, rather than just displaying images of current and projected views, the more the operator detects and corrects the error. Pacaux-Lemoine et al. conclude that an advanced interface that supports diagnosis activity for the operator's current activity should be used, so explanation, not merely illustration, is a key factor in understanding the manufacturing process.

Research has largely tended to focus on the different interfaces. The best choice of interface is based on the specific situation: that is a mix of the operators, the assembly task at hand, and the manufacturing environment [5]. There remains a research gap around user-centered considerations which are critical to system success. Efforts must be made to decrease the cognitive load by not providing any unnecessary distractions and using clear displays [18]. In order to reduce the occurrence of error, operator tasks must be made easier, simpler, and compatible with the capabilities of the operator [19].

Almost all of the papers refer to the need to develop a working system that integrates with other systems. In this research, an integrated Assembly 4.0 System is developed, and evaluated with operators. The operator is augmented with just enough technology and information to support the operator in completing a high-quality assembly. This might be assembly guidance when they need it, say on an unfamiliar SOP, or when an error has been made. They are given just enough information, at the right time, and in the right place, as recommended by Johansson et al. [13].

3 Research Method

In this research, an exemplar Assembly 4.0 system is developed and applied to a number of assembly use cases. The exemplar system is further described in patent application number WO2022053719A1 Intelligent cognitive assistant system and method [20]. A simple assembly use-case is chosen for the experiments. In this section, the system is first described, and then the experiments are outlined.

3.1 A Data-Driven Cyber-Physical Assembly 4.0 System

The system was coded in Python and developed using the Django full-stack framework. The frontend was developed in React. The system is modular which allows for features to be turned on and off ensuring scalability and accessibility. The assembly is checked using a computer vision approach. Machine learning models are used to set a state control signal that controls the assembly process. The system sits on top of the existing production line and therefore does not interfere with the existing process. Not impacting the process is a key user requirement from Process Engineers who wish to avoid the need for revalidation of a line with the FDA. In Fig. 1, the workflow for the system is provided. The operator is identified and authenticated, and the process instructions appropriate for the work order are retrieved from connected database systems, and assembly guidance can begin. The operator is advised as to the appropriate SOP assembly instructions.

The computer vision system identifies that the correct parts are being used. This approach improves on the work of Rupprecht et al. [8] and Quint et al. [9] where the pick-by-light feature identifies the parts to be used. Pick-by-light assumes that the parts and tools are put back into the correct locations, thus the operator is constricted. The computer vision system also observes the assembly work being executed. Importantly, and in contrast to the HAR approach as recommended by Bertram et al. [7] and Quint et al. [9], the vision system does not check the activity of the operator but

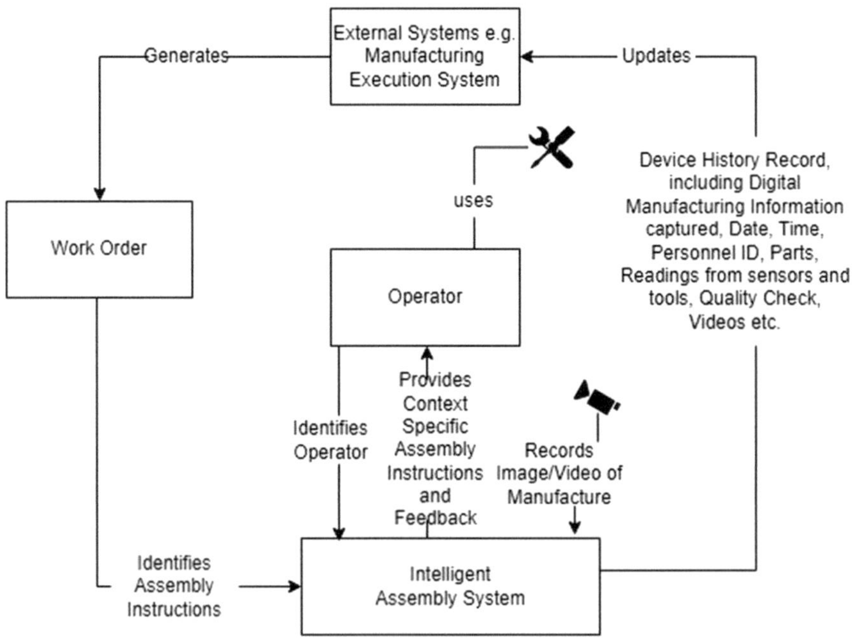

Fig. 1 Human-centric assembly 4.0 system

checks the result. This creates a situation where different approaches to the assembly are accommodated, consistent with the notion of Operator 4.0, as long as the part passes the quality inspection.

Quality inspection during the assembly process allows the system to recover from operator error [14]. In this system, the video feed and sensor data, such as measurement, are compared against trained machine learning models that verify that the work has been completed correctly. The sensor data is generated from metrology tools such as digital callipers, and digital torque wrenches. In multistage assembly, every part of the process has its own process parameters with an associated tolerance level, which can be measured by metrology, and evaluated within the system. Compliance within this tolerance level is deemed a quality pass.

Once the step has been verified, a system-wide trigger identifies that step as complete. The trigger sends a signal to the operator that the step was successfully completed. This signal can be provided using haptic or visual feedback. Should there be an error at this stage, the error is flagged to the operator, who is able to interpret, detect, and correct when mistakes occur. Additional control action is taken such as identifying the corrective steps and supporting the operator on the correct step. In this way, quality control is inline, and considered a step-by-step multistage reduction in variation from the process, and extends the single-stage quality assurance checks as recommended in Bakker et al. [21]. This inline automatic quality control check improves the works of Rupprecht et al. [8] which require the operator to signal by some action, such as pressing a button, or providing verbal feedback, that the work is correct.

Further, on step verification, a record is taken of that step, which includes all of the manufacturing data pertaining to the process. The data can be hosted in the cloud such as on an Amazon Web Service, or on a networked storage device.

The next step in the process sequence is then presented to the operator. The computer vision component can identify the step in the SOP that the operator is working on, thereby allowing for a non-linear process as may be used in load balancing, further improving on the work in Quint et al. [9].

The system provides information and guidance personalized to the preference and experience of the operator. As operators may be experienced on some SOPs and not on others, the amount of information and the choice of carrier of that information need not always be the same. Interfaces, information carriers, and communication are as simple as possible, to avoid a taxing cognitive load, are quick to consume, and do not interfere with the flow of work. The ultimate goal is for the system or SOP artifact to appear as 'simple' [2] by reducing the number of elements. To do this a Design for Simplicity (DfS) approach is used. The choice of information carrier is personalizable, so that a novice operator may choose to use video, aural, graphical, and/or textual information, whatever best suits the situation, as recommended in Wolfartsberger et al. [5]. These are multilingual.

In practice, operators do not seek assembly information when they believe the actions required are clear [1]. For this reason, the system can also operate in 'silent partner' mode. In 'silent partner' mode no assembly instructions are given but the system performs the inline quality checks at each step in the assembly process, and only flags the result to the operator when it believes an error might have occurred. In this way, the operator's attention is drawn to the particular part and step in the process.

A scenario of use is shown in Fig. 2. A new operator is building a catheter, and at this workstation eight steps in that process are completed. The operator logs onto the system and scans the work order. The system loads the appropriate process instructions and displays the first step of the process to the operator through their preferred feedback mechanism—in this case video display. The video instruction is a recording of an experienced operator performing the task. The video will loop until the operator has completed that step correctly Fig. 2(1). As the operator works on the process, the work is monitored through a live feed to cameras. This live feed is evaluated against a pre-trained model of the correct work Fig. 2(2). Multiple cameras allow for a three-dimensional representation of the scene. Once the AI determines the operator has completed the work correctly, the operator receives positive feedback through a signal to indicate a 'quality pass'. In this instance, Fig. 2(3), a green light is used to indicate the quality pass. The next steps in the process are loaded sequentially until all eight are complete, when the system will loop back to the beginning, and repeat, until the quantity required by the work-order has been completed. An Electronic Device History Record (eDHR) video record of the process is captured in Fig. 2(4).

The system was successfully tested on a number of complex use cases, such as the assembly of two different catheters, a stent, a starter motor for a car, a windscreen washer tank, and a twelve-plate mold for polymer injection molding.

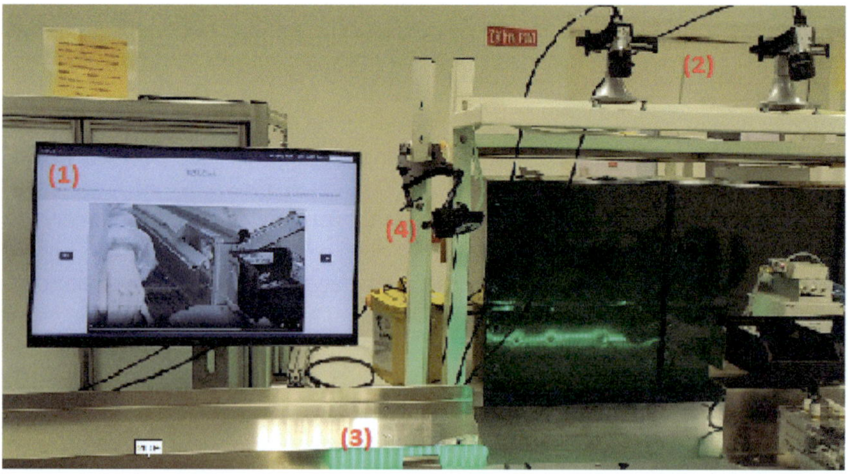

Fig. 2 Assembly system on line

3.2 Evaluation Method

The goal is not to measure the operational benefits of the system, which will form the basis of other studies, but to measure the potential technological adoption of the proposed solution by the people who are most impacted by the system. There are two experiments—one that seeks the operators' opinion, and the second the Process Managers. Questionnaire research features the use of self-report measures on selected questions and is an appropriate measurement instrument to capture subjective perspectives.

Experiment I: Evaluate the Operators' Perspective

The first experiment is an evaluation of the operators' perspective which tests for usability. The System Usability Scale (SUS) is highly reliable for evaluating subjective opinions on usability over a number of dimensions [22, 23]. In a review of 2,324 surveys over 206 studies, SUS was determined to give statistically significant results and is robust in assessing the quality of a broad spectrum of systems [24]. SUS uses a five-point Likert agreement scale, thus providing quantitative data for evaluation. SUS is composed of ten statements, which in order to avoid bias alternate between positive and negative [23]. The questions are listed in the leftmost column of Table 1.

Thirty operators participated in the study (N = 30), which is comparable with the sample size used in related studies [8, 11, 16]. The operators are provided with training on how to use the system. The operators are asked to complete the assembly, which is repeated five times, by following the guidance provided by the system. The product to be assembled was a pen. The pen was chosen so as to remove any

Table 1 Survey questions and sample response from the operator

Data field	Response
Observation number	1
SUS 1: I think that I would like to use this system frequently	5
SUS 2: I found the system unnecessary complex	1
SUS 3: I thought the system was easy to use	5
SUS 4: I think that I would need the support of a technical person to be able to use this system	1
SUS 5: I found the various functions in this system were well integrated	4
SUS 6: I thought there was too much inconsistency in this system	2
SUS 7: I would imagine that most people would learn to use this system very quickly	5
SUS 8: I found the system very cumbersome to use	1
SUS 9: I felt very confident using the system	4
SUS 10: I need to learn a lot of things before I could get going with this system	1
On a scale of 1–7, how successful do you think this system will be in reducing human error in assembly?	6

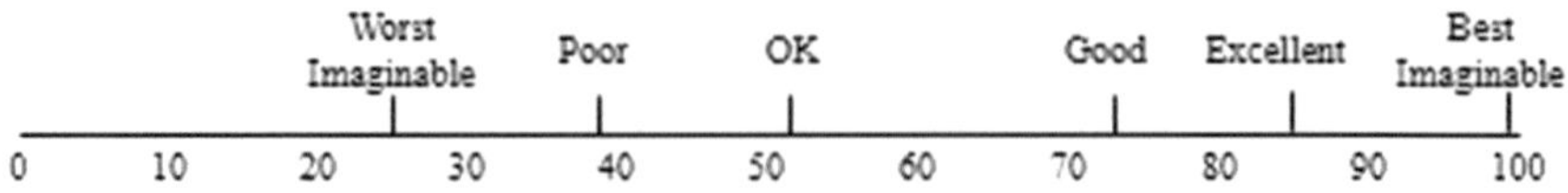

Fig. 3 System usability scale with adjective ratings

task complexity, which has an impact on mental demand, physical demand, temporal demand, effort, performance, and frustration levels. Task complexity may influence the participants' experience, and be reflected in ratings with respect to usability. The pen assembly comprises four steps: insert the thrust device the right way up into the barrel, insert the ink chamber into the barrel, add the spring to the ballpoint tip of the ink chamber, and screw the tip onto the barrel.

Post assembly the operators complete the questionnaire. Following this, the SUS scores are calculated. A sample set of data is listed in the 'Response' column in Table 1. The output from SUS is a single score that captures the participants' view of the system's usability. Final scores for the SUS can range from 0 to 100, where higher scores indicate better usability. According to Bangor et al. [24] a score of 70 indicates a passable system, scores in the high 70 s to upper 80 s indicate higher quality systems, and scores higher than 90 indicate truly superior systems. The overall SUS score collected in this experiment has a median value at the 87.5th percentile and a mean at the 82.42th percentile.

The adjective rating [22] shown in Fig. 3 is used. This scale is a six-point adjective rating scale that helps in the interpretation of the score. The adjective rating illustrates the Grade Scale, with an adjective mapping ranging from Worst Imaginable, Poor, OK, Good, Excellent, and Best Imaginable. The mean SUS score of 82.42, deems it to be a higher quality system, with an adjective rating just below "excellent."

Experiment II: Evaluate the Managerial Perspective

The second experiment is an evaluation from the Process Managers' perspective. The system was demonstrated at the Manufacturing Solutions Exhibition in June 2022 and in June 2023. The study participants ($N = 30$) were Process Engineers, who had responsibility for selecting, and managing new systems and tools to assist in achieving quality production. Participants were recruited from the attendees at the event. Their attendance at the event indicated an interest in solutions to production problems. Potential participants were invited to the stand at which the system was being demonstrated. They were screened based on a general discussion about the system and their roles at organizations.

The system was demonstrated to the participants on a range of assemblies. The participants were invited to use the system. Following the demonstration, the participants were asked for their feedback. The survey featured the questions listed in the leftmost column of Table 2. The responses were captured on a tablet in survey form and later exported as a spreadsheet. Thirty responses were evaluated for eligibility. Two responses were excluded from the results, as the respondents reported that they had no involvement or experience in the management of assembly that involved humans, leaving twenty-eight confirmed eligible and included.

Table 2 Managers' survey sample data

Data field	Response
On a scale of 1–7, rate the importance of reducing error in assembly by human operators	6
On a scale of 1–7, how easy is it to use this system?	6
On a scale of 1–7, how successful do you think this system will be in reducing human error in assembly?	7
Do you have any other comments?	May suit mid production output. We have two problems, first humans make mistakes. Second they forget the make them. The first problem you cannot solve, but this could help the second

4 Results and Discussion

The research question, will an Assembly 4.0 System be acceptable to the workers using the system, has been answered. The Assembly 4.0 system developed herein is based on the principles of Quality 4.0 which has at its core three pillars of process, human, and technology [25]. The human depends upon quality and context-specific process instructions. The human is augmented with the appropriate technology to complete the complex manufacture to a high standard of quality. Technology includes cyber-physical devices tools and systems, machine learning models, and vertical and horizontal integration with other systems. The trusted relationship between the human and the technology is evidenced in that the technology does not monitor the operator in a supervisor role but in a partner mode, flagging to the operator where attention needs to be brought. In this way it empowers the operator to take back ownership of the assembly process, fitting with the notion of Operator 4.0.

The first part of Experiment II is to identify whether the Process Managers believe that is there a need for an Assembly 4.0 System. Given the innovations within the industry, does the human error still cause pain points for the managers in charge of assembly, thus qualifying the need for this research? The importance of reducing human errors in assembly was reported by 100% of respondents with ratings in the upper part of the scale, of which 56% rate it at the top of the scale. Further, some of the qualitative statements corroborate this data. One manager stated "*Most companies are running around with their hair on fire. This will help*". Another stated "*Really interesting concept, it will cut down pressure on quality control who are under so much pressure, even if only cut down on work they have to do. There's also a business case as quality control is expensive to hire, and run*". So, human error is still a cause of concern, in a business under pressure, and therefore the need for Assembly 4.0 systems is apparent.

The satisfaction with the system by operators is evidenced in the SUS score [16]. This AS4.0 system was rated on the 82.42th percentile, which using the adjective score is classed as between very good and excellent.

There is a consistency between the operators and the managers' responses with regard to how easy it is to use this system. The operators rated the system as being easy to use, with 90% rating it in the two highest categories. There was one outlier scoring 2 out of 5. However, this 2 is not consistent with that participant's other scores, who scored 4/5 for "I think that I would like to use this system frequently," and 5/5 for "I would imagine that most people would learn to use this system very quickly," and 5/5 for "I felt very confident using the system." The managers rated it as being easy to use, with 100% ranking it in the upper limits of the scale, with 78% rating it in the two highest categories. While it appears that it is easy to use and acceptable from a user satisfaction perspective, there are insights into potential acceptance issues from the qualitative data. The following comments were made "*great system, but need to be careful with unions*" and another who worried "*about the information being stored and question unions acceptance of such a system*", and another "*problem will be operator buy in*" who advised "*needs a process for successful adoption*". The potential for AS4.0 to help shift the culture in assembly is seen in the comment "*There is a dumbing down of work in manufacturing which leads to high turnover. If this system helps the operator then it would be great.*"

With respect to ensuring acceptance with the operators, it will be necessary to have "*round table with the operators and unions*" and ensure that there is compliance, as "*GDPR will be an issue as identifying the person*". There is an issue of adoption resistance by operators in the use of human-cyber-physical systems [26], particularly when data is being recorded for monitoring [27]. In order to ensure acceptance, operators and other workers need to continue to be part of the participatory design, and the requirements of the system stay true to the concept that the system augments the operator, keeping the operator central to the process. If AS4.0 becomes another monitoring tool, it feeds into the view of treating the operator as a risk, which goes against the principles of Operator 4.0.

The research gathered the subjective opinion of both groups of workers as to their belief that such a system will have a successful impact on the reduction of human error, using a scale of 1–7. The data are non-parametric, and have homogeneity of variances (Levenes Test $F = 1.1811$, $p = 0.2819$) so a Wilcoxon Rank Sum test is conducted on the two independent samples. There is no statistical difference between the two groups of workers (Wilcoxon Rank Sum $Z = 1.19551$, $p = 0.2319$), with a mean of 5.6 for operators, and 5.9 for Process Managers, and a median of 6 for both groups.

Having said that the experiments herein bode well for adoption success. The managers' score is very positive, and the majority of comments were positive "*great system*", "*very simple interface, easy to use*", "*very good product*", "*very interesting*". These managers review tools and make the purchasing decisions for their operators. The operators scored the product highly, which bodes well from their perspective. Operators can choose to use the tool in a meaningful way thereby making the most of its features and functionality.

5 Conclusion

A new data-driven cyber-physical AS4.0 system is described. The system provides Intelligence Augmentation to the operator, through the provision of inline quality check data, simple access to the SOP, and easy access to the experienced operator's understanding of the process. The system's design is rooted in the current schools of thought around Operator 4.0 and Quality 4.0. With respect to Operator 4.0, the system keeps the human central to the assembly process, and in doing so leverages their skill, gut feeling, experience, and intuition as to how assembly should be completed. With respect to Quality 4.0, the system is built on three pillars—human, technology, and process. Context-sensitive process support is given to the operator. The system is designed in best practices with respect to modularity of code, cloud based elastic compute, is cyber secure and integrates with IoT and external systems. The issue of operators not looking at assembly instructions is addressed by the use of an intelligent assistant which provides context specific help should the operator require it, cognizant of the fact that the operator should not need constantly to refer to assembly instructions, but due to humans factors, such as fatigue, can occasionally make an error. This system supports the operator in those moments to draw their attention to it, and in doing so augments the operator with just the right information to complete the assembly without error. In this way the operator becomes their own quality assessor, being empowered by AS4.0, to identify the error, and rework the same before it is propagated throughout the system. The system is evaluated by operators who score it very favorably from a usability perspective. Usability is key to technology acceptance and bodes well for adoption.

The system is evaluated by process managers, who score it highly as easy to use, and its likely impact on reducing human error. They also validate the need for the research by confirming that human error is still a significant problem, for which systems such as this can help. Insights into the successful implementation of AS4.0 were gathered from the qualitative data. The system was declared "great", "interesting" and so on, but warnings are given for the need for operator buy-in and to ensure that the system does not monitor the operator in a supervisory role. The goal of AS4.0 grounded in the notions of Operator 4.0 and Quality 4.0 is that the operator is the supervisor or 'magic human' augmented by technology to assist them in their work, rather than the other way around, keeping this as a fundamental user requirement in the development of future AS4.0 is essential.

5.1 Future Work

While the positive reception from the workers, reported here, should bode well for adoption success, future work should measure the operational benefits of the system. Empirical evidence is needed to consider the value, such as on the reduction of the number of human assembly errors made, the number of errors caught and reworked during the assembly, and the impact on quality.

Acknowledgements This research is co-funded by the Enterprise Ireland and European Regional Development Fund (ERDF) under Ireland's European Structural and Investment Funds (ESI) Programmes 2014–2020. Appreciation is given to R.Berry, I.Ashad, J.Scully, D.Byrne, and P.Quigley who assisted in the experiment.

References

1. Parmentier DD, Van Acker BB, Detand J, Saldien J (2020) Design for assembly meaning: a framework for designers to design products that support operator cognition during the assembly process. Cogn Technol Work 22(3):615–632. https://doi.org/10.1007/s10111-019-00588-x
2. ElMaraghy W, ElMaraghy H, Tomiyama T, Monostori L (2012) Complexity in engineering design and manufacturing. CIRP Ann 61(2):793–814. https://doi.org/10.1016/j.cirp.2012.05.001
3. Maettig B, Foot H (2020) Approach to improving training of human workers in industrial applications through the use of intelligence augmentation and human-in-the-loop. In 2020 15th international conference on computer science and education (ICCSE). IEEE, Aug 2020, pp 283–288. https://doi.org/10.1109/ICCSE49874.2020.9201867
4. Moore S (2011) A multidimensional visualisation model for zero defect in a biomechanical manufacturing environment. University of Limerick
5. Wolfartsberger J, Haslwanter J, Lindorfer R (2019) Perspectives on assistive systems for manual assembly tasks in industry. Technologies 7(1):12. https://doi.org/10.3390/technologies7010012
6. Cohen Y, Naseraldin H, Chaudhuri A, Pilati F (2019) Assembly systems in industry 4.0 era: a road map to understand assembly 4.0. Int J Adv Manuf Technol 105(9):4037–4054. https://doi.org/10.1007/s00170-019-04203-1
7. Bertram P, Motsch W, Rübel P, Ruskowski M (2019) Intelligent material supply supporting assistive systems for manual working stations. Procedia Manuf 38:983–990. https://doi.org/10.1016/j.promfg.2020.01.182
8. Rupprecht B, Trunzer E, Kovac J, Vogel-Heuser B (2020) Conception and development of a support system for assembly technology, pp 639–657. https://doi.org/10.1007/978-3-030-49065-2_44
9. Quint F, Loch F, Orfgen M, Zuehlke D (2016) A system architecture for assistance in manual tasks. IOS Press. https://doi.org/10.3233/978-1-61499-690-3-43
10. Gürerk Ö, Bohné TM, Alvarez Alonso G (2018) Productivity and learning effects of head-mounted AR displays on human-centered work. SSRN Electron J. https://doi.org/10.2139/ssrn.3264118
11. Pacaux-Lemoine MP, Poulain T, Debernard S (2013) Human-machine cooperation and manufacturing system. In: IECON 2013—39th annual conference of the IEEE industrial electronics society. IEEE, Nov 2013, pp 8288–8293. https://doi.org/10.1109/IECON.2013.6700521
12. Fässberg T, Fasth Å, Stahre J (2012) A classification of carrier and content of information. In: Proceedings of the 4th CIRP conference assembly technologies systems, pp 1–4. http://cirp.me.engin.umich.edu/CIRP_CATS_2012_Proceedings_small.pdf
13. Johansson PEC, Malmsköld L, Fast-Berglund Å, Moestam L (2018) Enhancing future assembly information systems—putting theory into practice. Procedia Manuf 17:491–498. https://doi.org/10.1016/j.promfg.2018.10.088
14. Torres Y, Nadeau S, Landau K (2021) Classification and quantification of human error in manufacturing: a case study in complex manual assembly. Appl Sci 11(2):749. https://doi.org/10.3390/app11020749

15. Bortolini M, Ferrari E, Gamberi M, Pilati F, Faccio M (2017) Assembly system design in the Industry 4.0 era: a general framework. IFAC-PapersOnLine 50(1):5700–5705. https://doi.org/10.1016/j.ifacol.2017.08.1121
16. Havard V, Baudry D, Jeanne B, Louis A, Savatier X (2021) A use case study comparing augmented reality (AR) and electronic document-based maintenance instructions considering tasks complexity and operator competency level. Virtual Real 25(4):999–1014. https://doi.org/10.1007/s10055-020-00493-z
17. Borisov N, Weyers B, Kluge A (2018) Designing a human machine interface for quality assurance in car manufacturing: an attempt to address the 'functionality versus user experience contradiction' in professional production environments. Adv Human-Computer Interact 2018:1–18. https://doi.org/10.1155/2018/9502692
18. Mattsson S, Li D, Fast-Berglund Å (2018) Application of design principles for assembly instructions—evaluation of practitioner use. Procedia CIRP 76:42–47. https://doi.org/10.1016/j.procir.2018.02.011
19. Bhise VD (2013) Human factors engineering in product design. In: Designing complex products with systems engineering processes and techniques. CRC Press, pp 218–237. https://doi.org/10.1201/b15483-15
20. Humphries J (2021) Intelligent cognitive assistant system and method. WO2022053719A1. https://patents.google.com/patent/WO2022053719A1
21. Bakker OJ, Ratchev SM, Popov AA (2015) Towards a condition-monitoring framework for quality assurance in intelligent multistage manufacturing Environment. IFAC-PapersOnLine 48(3):2089–2094. https://doi.org/10.1016/j.ifacol.2015.06.397
22. Bangor A, Kortum P, Miller J (2009) Determining what individual SUS scores mean: adding an adjective rating scale. J Usability Stud 4(3):114–123
23. Klack L, Schmitz-Rode T, Wilkowska W, Kasugai K, Heidrich F, Ziefle M (2011) Integrated home monitoring and compliance optimization for patients with mechanical circulatory support devices. Ann Biomed Eng 39(12):2911–2921. https://doi.org/10.1007/s10439-011-0407-1
24. Bangor A, Kortum PT, Miller JT (2008) An empirical evaluation of the system usability scale. Int J Human-Computer Interact 24(6):574–594. https://doi.org/10.1080/10447310802205776
25. American Society for Quality (ASQ) (2022) QUALITY 4.0. https://asq.org/quality-resources/quality-4-0. Accessed 02 Apr 2022
26. Kamble SS, Gunasekaran A, Gawankar SA (2018) Sustainable Industry 4.0 framework: a systematic literature review identifying the current trends and future perspectives. Process Saf Environ Prot 117:408–425. https://doi.org/10.1016/j.psep.2018.05.009
27. Kaasinen E et al (2020) Empowering and engaging industrial workers with operator 4.0 solutions. Comput Ind Eng 139:105678. https://doi.org/10.1016/j.cie.2019.01.052

Author Index

G. L. Conte and O. Sename (eds.), *Proceedings of 12th International Conference on Mechatronics and Control Engineering*, Lecture Notes in Mechanical Engineering,
https://doi.org/10.1007/978-981-96-6452-8

MIX
Papier aus verantwortungsvollen Quellen
Paper from responsible sources
FSC® C105338

If you have any concerns about our products,
you can contact us on
ProductSafety@springernature.com

In case Publisher is established outside the EU,
the EU authorized representative is:
Springer Nature Customer Service Center GmbH
Europaplatz 3, 69115 Heidelberg, Germany

Printed by Libri Plureos GmbH
in Hamburg, Germany